WITHOUT EXPERIMENTATION, A WILLINGNESS TO ASK QUESTIONS AND TRY NEW THINGS, WE SHALL SURELY BECOME STATIC, REPETITIVE, AND MORIBUND.

— Anthony Bourdain

Salty Walt's NEXT LEVEL Portable Antenna Sketchbook

by Walt Hudson, K4OGO

Published by
ARRL The National Association for Amateur Radio®
225 Main Street, Newington, CT 06111-1400 USA
www.arrl.org

Printed in the USA

ISBN: 978-1-62595-237-0

First Edition, First Printing

Photograph Credit:
All photographs that do not have a source identification courtesy of Walt Hudson, K4OGO.

Editorial Team:
Acquisitions Editor
Mark Derks, KC1RVQ

Book Editor
Makenzie Ozycz

Production by:
Terese Newman, KC1YPP

Salty Walt Illustration by:
Gabrielle McKay

We strive to produce books without errors. Sometimes mistakes do occur, however. When we become aware of problems in our books (other than obvious typographical errors), we post corrections on the ARRL website.

If you think you have found an error, please check **www.arrl.org/product-notes** for corrections. If you don't find a correction there, please let us know by sending an email to **pubsfdbk@arrl.org**.

Contents

Next Level Antenna Build Projects

Next Level "Off the Shelf" Antennas

Preface

Some say that a musician's best songs are found on their second album. Maybe that's because they continued to learn and improve their craft. For me, this book is that second album: a place to share even better antenna building songs. In my first book, *Salty Walt's Portable Antenna Sketchbook*, I shared my passion for investigating, sketching, and building portable ham radio antennas hoping that it would inspire others to grab some wire and make some contacts. Those antennas were mostly simple vertical, horizontal, and sloping wire antennas that are relatively easy to build and yield good results for portable HF ham radio outings.

This book is the next step. Experimenting and learning by either building or using antennas that are on the "Next Level." This collection provides an introduction to beam forming and directing HF radio signals. It discusses increasing an antenna's gain, understanding its far field pattern and take-off angle, and getting the most efficiency from an antenna based on its setup and application.

I've built or used all these antennas and more on my YouTube channel "Coastal Waves & Wires." You can watch me putting these antennas up and using them to make contacts, but the only way to really learn and enjoy the amateur radio hobby is to get out there and do it yourself. I hope these books give you the inspiration to do that.

I stated in my last book that, "if you want the Mozart level symphonic guide to antennas, this is not the book." This is the Jimmy Buffett of antenna books. Well, this book is the second album. The songs are only getting better.

73, Salty Walt, K4OGO

Introduction – Next Level Portable Antennas

I love building antennas, buying antennas, modifying antennas, and most of all, using them out portable on the HF bands. In this day and age of great SDR transceivers, inexpensive QRP transceivers, and affordable used transceivers, it seems that homebrew radios and kit building are just not as popular as they once were, and antenna building is the biggest and most popular way for amateurs to experiment and build projects in our hobby.

I'm a DXer who operates with antennas mostly for the HF bands. This book is a continuation of the collection of my sketches, thoughts, ideas, and experiences from using those antennas. This second book is another season in my research and experimentation with antennas. It records my research and experiences with HF antennas that I consider to be on the next level for portable ham radio operators and enthusiasts.

With that said, let's go have some fun. Let's put up some more antennas and experiment. That's what amateur radio is all about, and I hope this book encourages you to get out and do just that. These antennas are more of a challenge, more of an effort to build. Because they require more effort to put together, they can also be more frustrating when they don't work exactly as planned. Keep going! Making mistakes, overcoming them, and putting something awesome together is the absolute best way to master and learn anything.

Throughout the book I'll give you an overview of each antenna along with my modelling findings and my experience using it. I'll list the items needed to build the antenna and then my sketch. As with my first book, you don't have to read the instructions. You can just look at the pictures, but reading the instructions will increase your chance of a successful build.

This is the age of great radios. I love small, portable HF transceivers, and I have a collection of new and old radios I take out on the beaches and to the parks to play with. At the end of each chapter I'll share my collection and give my thoughts on these rigs.

Let's go have some fun!

Next Level
Antenna Build Projects

THE THREE-ELEMENT VERTICAL BEAM

DX COMMANDER

The three element Yagi-Uda antenna in its simplest form is the basic beam antenna that all beams are designed and built upon. I've done this vertically using fiberglass poles and wire. Before we look at my vertical version of this antenna, let's understand what a three-element Yagi antenna is.

A three-element Yagi-Uda antenna (also called a beam antenna or parasitic array) is a directional antenna consisting of one driven element and two parasitic elements (a reflector and a director). It is extremely popular because of its balance of high gain, compact size, and relatively simple construction.

Core Components

Driven Element: The central element connected directly to the transmission line. It is typically a half-wave dipole.
Reflector: The longest element, placed behind the driven element. It "reflects" energy forward, increasing gain in the desired direction.
Director: The shortest element, placed in front of the driven element. It "directs" the signal to sharpen the beam.

Typical Specifications

Gain: Generally provides 5 to 8.5 dBd (7 to 10.6 dBi) of gain compared to a standard dipole.
Front-to-Back Ratio: Offers significantly better directional performance than a 2-element Yagi, with ratios typically between 15 and 25 dB.
Spacing: Elements are usually spaced approximately 0.15 to 0.20 wavelengths apart.
Element Lengths: As a general rule, the reflector is 5% longer and the director is 5% shorter than the driven element.

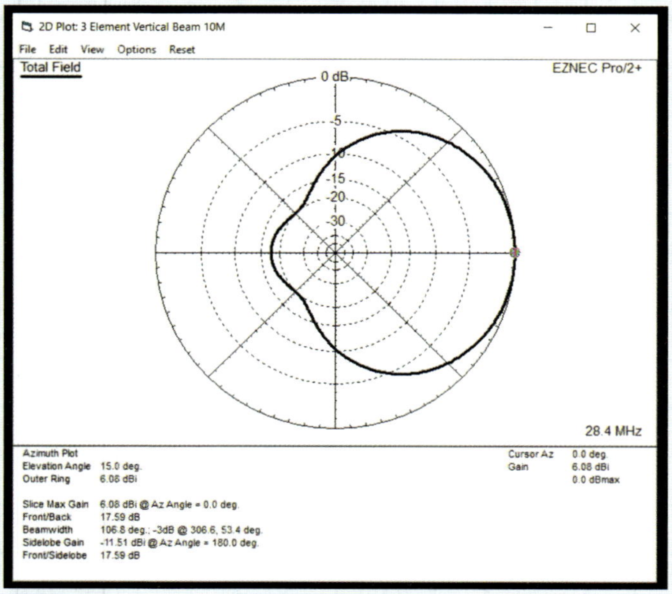

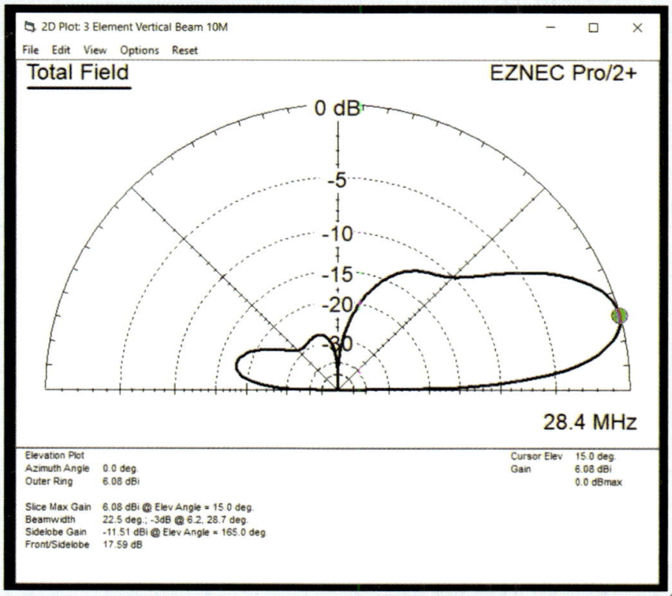

The Yagi-Uda antenna was invented by Shintaro Uda and Hidetsugu Yagi of Tohoku Imperial University, Japan in 1926 and is now often referred to as just the "Yagi."

This antenna is extremely popular in the VHF and UHF world of RF transmissions. For those of us that grew up before cable and satellite TV, we remember almost every house in our neighborhood having an antenna on top of it for television reception; many of those were Yagis!

Yagis can be very large on the amateur radio HF bands, making them tough to consider "portable antennas," but I have found that it's not impossible at all to build for the higher HF bands, especially 10 meters. I built this antenna for the 10-meter band with excellent results using three 7-meter-tall DX Commander poles, some 16-gauge wire, and a 1:1 balun at the feed point.

Getting this up was an easy task as I just took three pieces of small PVC pipe for the base, decided on a direction to aim my beam, and measured out the spacing between each element. Then it was just a matter of sliding the wires that I taped to the poles on the PVC base. Changing direction of the antenna was somewhat easy as well, by just swapping the director and reflector I could aim 180 degrees in the opposite direction.

This antenna could be built for bands lower than 10 meters by using larger poles but may be more of a challenge to erect, especially for one person alone, but don't let me discourage you from trying! Do the math. Remember that the driven element in the center is a resonant vertical dipole, the director in front is 5 percent smaller and the reflector behind it is 5 percent longer. Center the wires and space them between 0.15 and 0.20 wavelengths apart (I found around 0.17 to be the sweet spot and a great starting point). If you would rather just build the 10-meter version without having to do too much scratching of the brain, check out my sketch and build it as I did. It's an awesome antenna.

3 ELEMENT VERTICAL BEAM

FOR 10 METERS

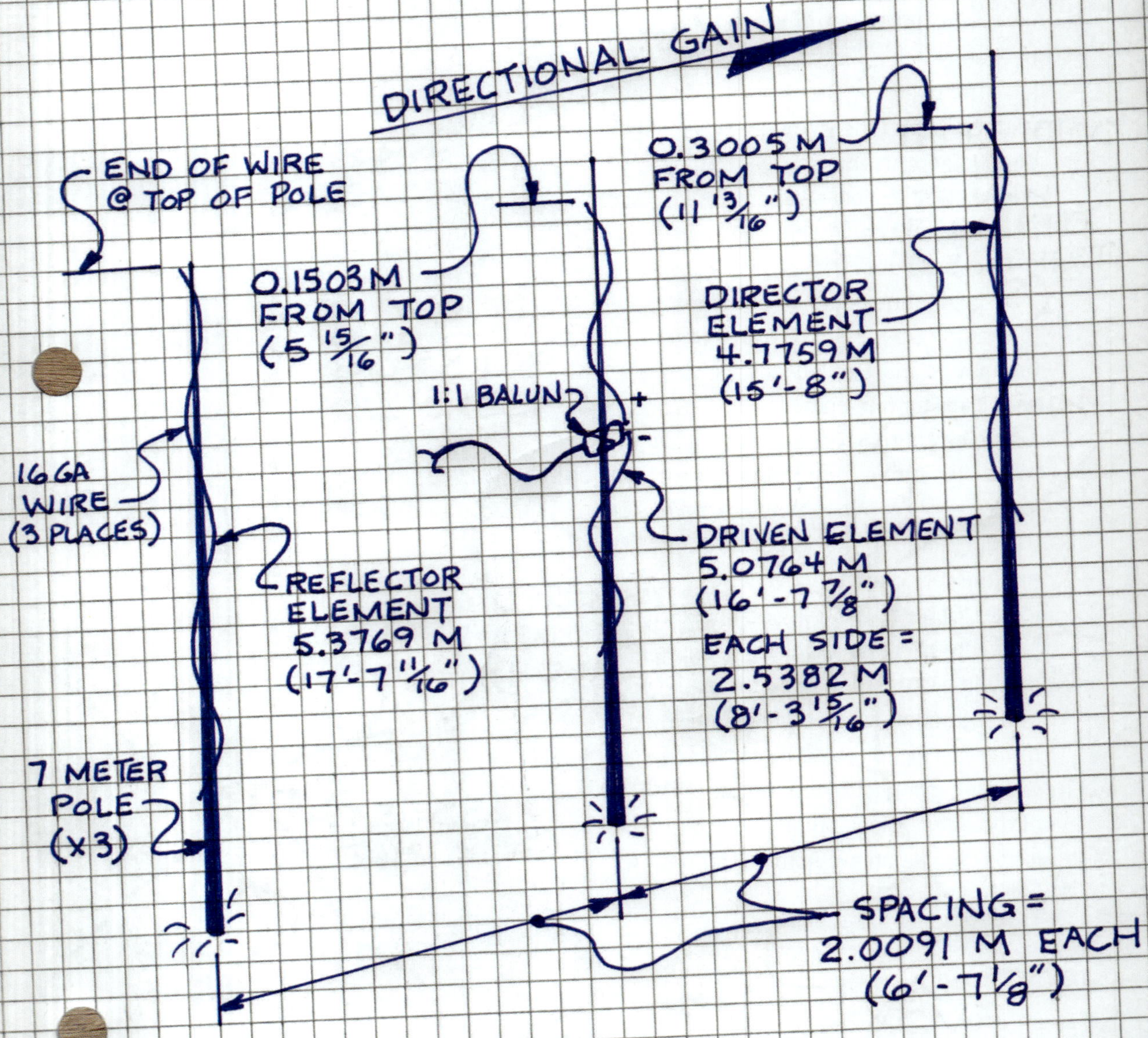

SALTY WALT K4OGO

Salty Walt's Portable Rigs

Xiegu G90

As is only fitting, the first portable radio in this book — and probably my all-time favorite — is the **Xiegu G90**. This amazing little, inexpensive transceiver does it all and is always the radio I suggest to every new ham when they're getting started in the portable HF side of our hobby. It only packs 20 watts, but that's more than enough to make successful contacts around the world portable. It's a software-defined radio (SDR for short) and is the entry level, most affordable way to have a waterfall screen while portable. The waterfall lets you see all of the band and search for contacts while hunting signals. This rig also has great antenna analysis tools built in that allow you to check SWR and adjust any antenna setup without needing a separate analyzer. But all that is just a small part of what makes this radio so good. It has — without a doubt — the absolute best internal antenna tuner (ATU) built into it and will tune absolutely anything, which is great for anyone just getting started in building portable antennas. If that's you, there is no better radio to start with than the **Xiegu G90**.

The Two-Element Quarter-Wave Vertical with a Reflector

Using the lessons learned from the three-element Yagi-Uda antenna, it's possible to turn a simple quarter-wave ground plane antenna into a two-element beam antenna. In its simplest form a quarter-wave vertical ground plane antenna is basically a vertical dipole that uses the ground to mirror one-half of the antenna. Why not just put a reflector behind it that also uses the ground as a mirror?

I thought this would be easy to do if I used two popular 17-foot telescopic whip antennas that can be tuned by raising or lowering the elements. I positioned them so that the antenna system was resonant for any band between 20 and 10 meters. One very important and key component to the antenna system is a jumper or wire between the upper element and the ground radials on the back reflector to make it mimic one complete run of wire.

So, a quarter-wave vertical antenna with a reflector is essentially a two-element vertical parasitic array. While a standard quarter-wave vertical is omnidirectional, adding a parasitic reflector behind the vertical converts it into a directional antenna with forward gain.

Core Components

Spacing: The reflector is placed approximately 0.15 to 0.25 wavelengths behind the driven vertical element.

Sizing: To act as a reflector, the parasitic element must be electrically longer than the driven element (typically by 2 to 3%).

Grounding: The reflector is not directly fed by the radio and the upper element is directly connected to the radial wires below it.

Typical Specifications

Gain: Adding a single reflector typically provides a gain of 3 to 4 dB in the forward direction compared to a standalone vertical.

Front-to-Back Ratio: The array offers a decent front-to-back ratio of 10.24 dB.

Directionality: The reflector suppresses signals from the rear and concentrates radiation in the forward direction (opposite the reflector).

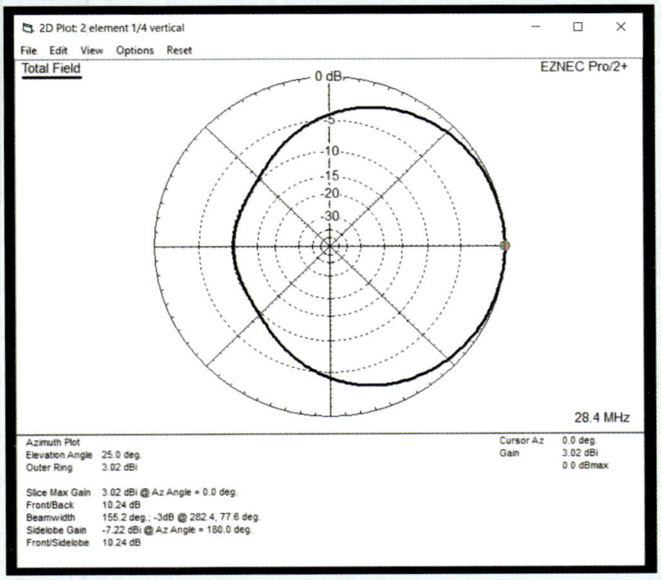

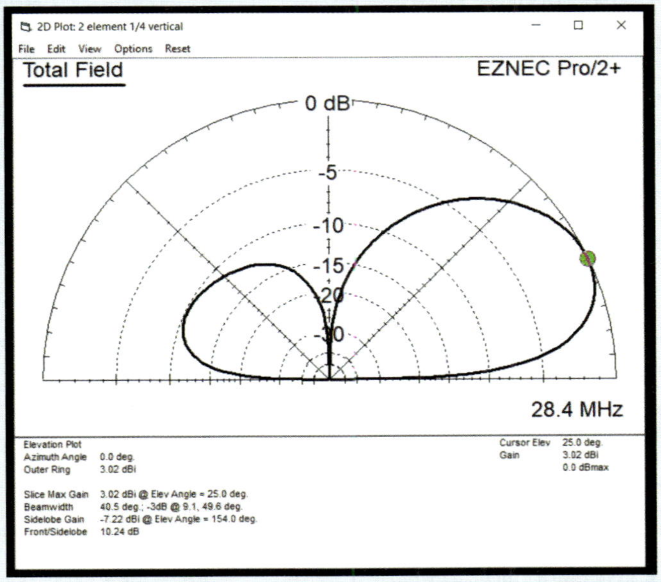

This is a very easy antenna to get started with when trying to get directional with a portable antenna, and can be reconfigured quickly and easily using just a measuring tape. This is also a great portable DX antenna as vertical antennas have a low take-off angle and adding a reflector focuses this energy on a specific target region.

One note, adding a parasitic element will lower the feed point impedance of the driven element. You may have to adjust and tune the driven element to get good SWR and make it resonant. When I did this, I made note of the amount of adjusting I did to the driven element and then slightly adjusted the reflector about the same amount. You don't have to be spot-on-perfect, but it helps to slightly adjust the reflector to suit.

Give this antenna a try. It could be your introduction to beamforming (focusing transmitted signals in a specific direction) and making directional antennas.

TELESCOPIC ANTENNA WITH A REFLECTOR

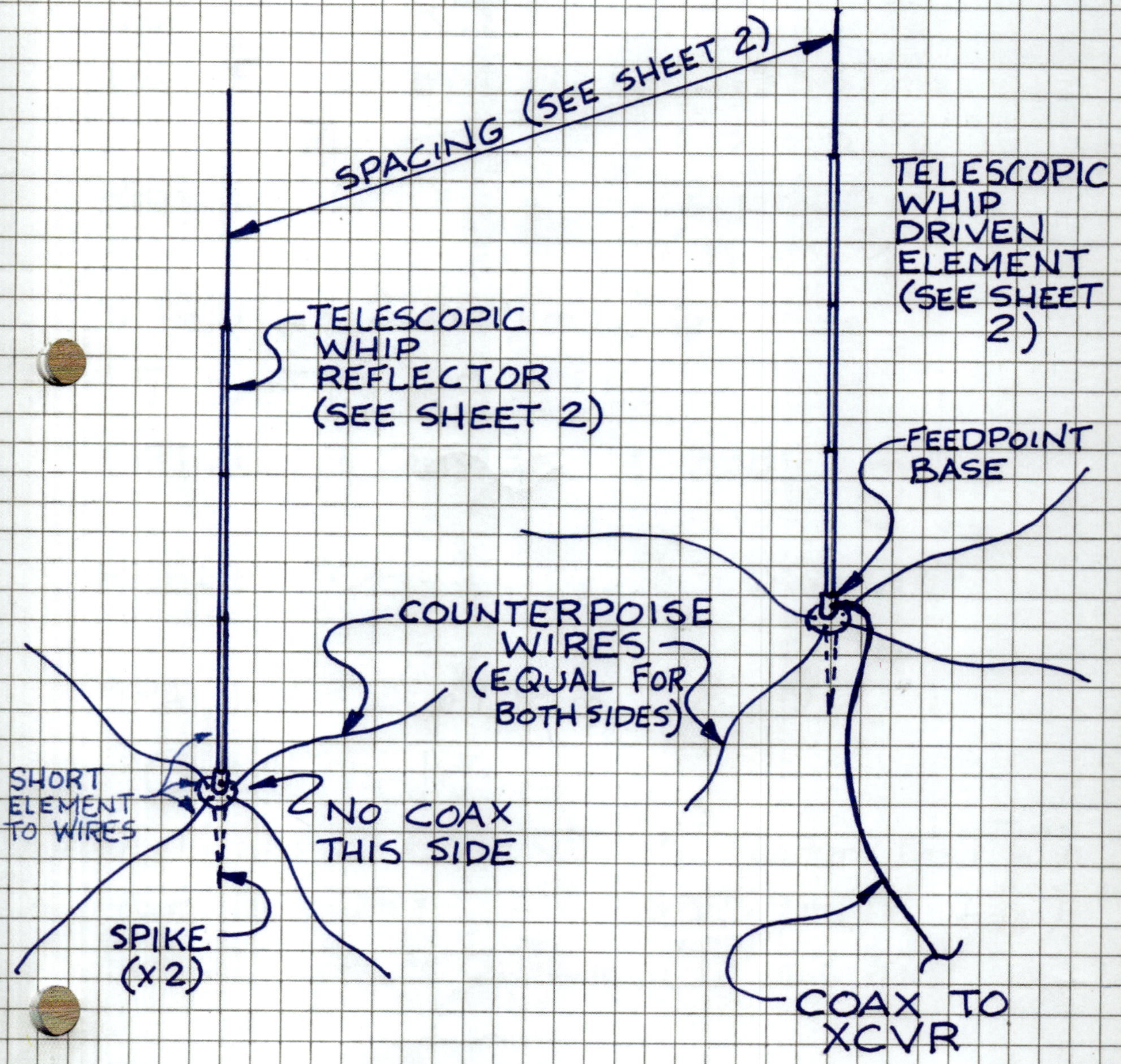

SHEET 1 OF 2

SALTY WALT K4OGO

TELESCOPIC WHIP EXTENSION LENGTHS & SPACING

10 METERS:

DRIVEN ELEMENT = 2.5382 m (8'-3 15/16")

REFLECTOR = 2.6885 m (8'-9 7/8")

SPACING = 2.0091 m (6'-7 1/8")

15 METERS:

DRIVEN ELEMENT = 3.3843 m (11'-1 1/4")

REFLECTOR = 3.5846 m (11'-9 1/8")

SPACING = 2.6788 m (8'-9 7/16")

20 METERS:

DRIVEN ELEMENT = 5.0764 m (16'-7 7/8")

REFLECTOR = 5.3769 m (17'-7 11/16")

SPACING = 4.0182 m (13'-2 3/16")

SHEET 2 OF 2

SALTY WALT K4OGO

YAESU
HF/50MHz TRANSCEIVER FT-710
MODE
SPLIT
CLAR
NB
LOCK
TUNE
PHONES
MIC
14.196.000
USB
LSB
7.000.000
CENTER
3DSS
MULTI
EXPAND
SPAN
SPEED
A/B
BAND
FUNC
DNR
BUSY/TX
NAR
STEP/MCH
DSP
DSP RESET
AF GAIN

Salty Walt's Portable Rigs

The FT-710

For the longest time I resisted going portable with a 100 watt transceiver. I thought to myself, I can get it all done with 20 watts or less. I was the typical pound-on-your-chest QRPer. Until I used this rig and fell in love with it. The Yaesu FT-710 Field is a compact, 100W HF/50MHz Software-Defined Radio (SDR) transceiver designed for portable operations, offering high-end receiver performance in a travel-friendly package. The "Field" version is identical to the "AESS" version internally, but it omits the big external SP-40 speaker and includes a carrying belt kit for enhanced portability. I personally don't use the belt kit, as I found that simply putting it in a tactical carry bag made transporting it out portable much easier.

This awesome radio covers the HF amateur radio bands (1.8 MHz to 50 MHz) and the 6-meter band, with 4-meter coverage available in the UK version. I actually ordered mine with the MARS mod which opens up all bands. It's power output is adjustable from 5W to 100W, it has a built-in tuner and only weighs 9.9 pounds (4.5 kg). The 4.3-inch color touchscreen is a big plus, and what I love most about this rig is how easy it is to operate out of the box. It's very menu friendly, unlike many other transceivers in this range that I have used.

A Two-Element Vertical Yagi with a Director

Using just a couple of fiberglass poles and some wire, you can build this antenna quickly and easily, and it provides great results. This antenna is truly a vertical "Yagi" antenna in its purest form. It consists of one driven element (a dipole) and one parasitic director element placed in front of it.

You might ask, why use a director over a reflector? Here are one advantage and disadvantage:

Higher gain: A director-driven Yagi can achieve up to 7 dBi of free-space gain, which is roughly 1 dB higher than a standard reflector-driven two-element Yagi.

Narrower bandwidth: The primary drawback is a significantly narrower operating bandwidth compared to reflector designs.

Core Components

Driven Element: A half-wave dipole for 10, 12, or 15 meters.

Director: A shortened element that directs the signal, increasing the antenna's gain. Usually 5% shorter than the driven element.

Typical Specifications

Gain: 4.6 dBi

Front-to-Back Ratio: 3 dB

Elevation Angle: 14°

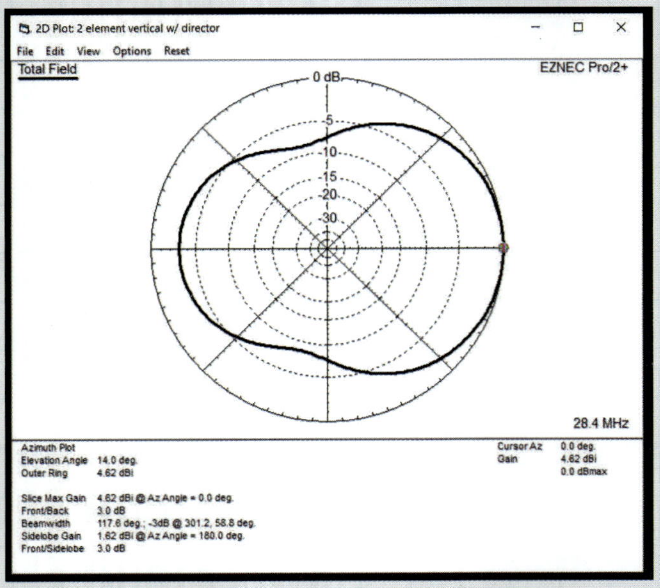

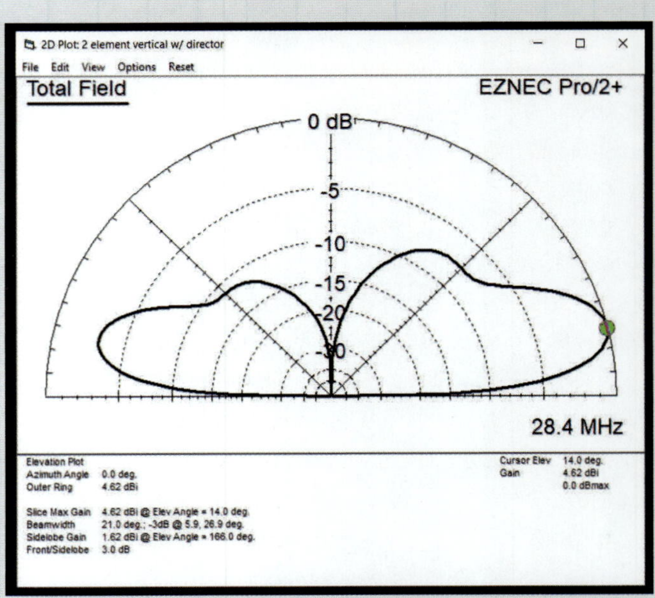

I personally have no problem using a tuner to overcome the narrow bandwidth issue, especially if I'm squeezing out a little more gain on the frequency I built the antenna for. The basic wire lengths and spacing for this antenna are the same as the three-element version mentioned in Chapter 1, and this antenna is buildable for the higher HF bands (15 through 10) using either 7 or 10 meter poles. I have built this antenna using my DX Commander poles by laying the two poles side-by-side on the ground and taping the wires to the pole to assure they are properly aligned, then driving two PVC pipes into the ground with proper spacing and aimed in the direction I desired, followed by sliding the poles onto the pipes. This also gives you two directions as you can simply switch the two poles to turn the beam 180 degrees. Another idea is to bring along several pieces of PVC pipe to support the antenna and place them around the driven element and then simply move the director around to suit the direction you want to beam your signal. It always feels good to say to someone on the air, "Let me turn my beam in your direction."

2 ELEMENT VERTICAL YAGI

ANTENNA WITH DIRECTOR

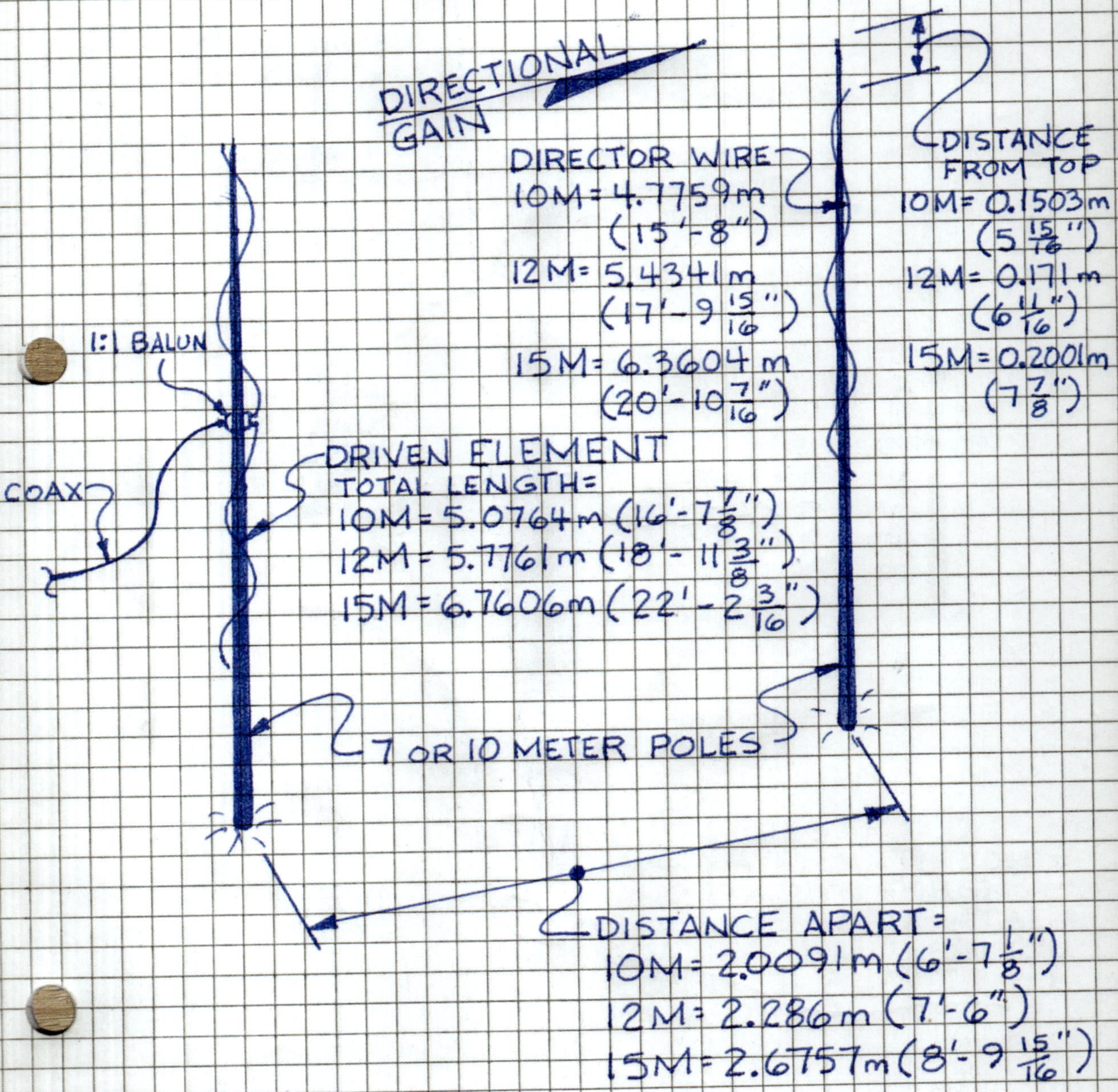

WALT K4OGO

Salty Walt's Portable Rigs

The QT80/AT-6666 PRO

The Radioddity **QT80** is a high-powered 10 meter mobile multimode transceiver that supports FM, AM, SSB, and CW. The **QT80** is identical to the Anytone **AT-6666 PRO**, and shouldn't be confused with the earlier AT-6666, which is an earlier model. The radio has 80 W peak envelope power (PEP) output and advanced noise reduction capabilities. In the radio world it is known as an "export" radio. This term is used for rigs that are marketed to hams for use on 10 meters but are popular in the CB world because they are easily modified to work on the 11 meter band. Of my entire collection of radios, this is the only one I have cracked open and modified, and no, it wasn't for the 11 meter capabilities. Modifying this radio also opens up the 15 and 12 meter bands, which makes it a great portable ham rig for the higher bands, giving me access to 15, 12, and 10. The mod was also a solderless fix that simply required me to move a jumper over on the board. I should note, Anytone's earlier model AT-6666 can't be modified. I love having the 80 watts portable, and this radio's audio and quality are well worth the price.

The Half Square

For pure portable DX operations, this antenna gives the most bang for the effort than any wire antenna I've ever built. As DXers, we have all made unexpected contacts and amazing long distance QSOs that kept us smiling long after hanging up the mic. This antenna gave me one of those experiences. Early one morning, I was portable with the half square making SSB voice contacts from coastal Virginia to Australia on 20 meters with solid S9 reports using 100 watts when suddenly a ZS call came through. He informed me that he was beaming long path to Australia and our QSO was definitely on the long path. That 26,628 kilometer (16,546 mile) contact with ZS5JY in South Africa brought goosebumps to my arms thanks to a half square antenna.

So, what exactly is a half square antenna? In its simplest form, it's two pieces of wire arranged as a phased array. That is, two quarter-wave vertical elements separated by one-half a wavelength of horizontal phasing line. To build this antenna, simply put a 1:1 balun up high on one pole and let a quarter wavelength of wire connected to it hang straight down. Connect a second wire that is three-quarters of a wavelength long to the balun and attach it horizontally to a second pole a half wavelength apart. Then bend the wire 90 degrees so that the remaining quarter wavelength of wire hangs down. This forms a second vertical element exactly one-half of a wavelength apart from the first wire. So, we are making a two-element phased array beam using two wires, two poles, and a 1:1 balun. It's that simple and easy.

Core Components

Wires: One ¾-wavelength wire and one ¼-wavelength wire.

Balun: 1:1

Supports: Two 7- or 10-meter fiberglass poles.

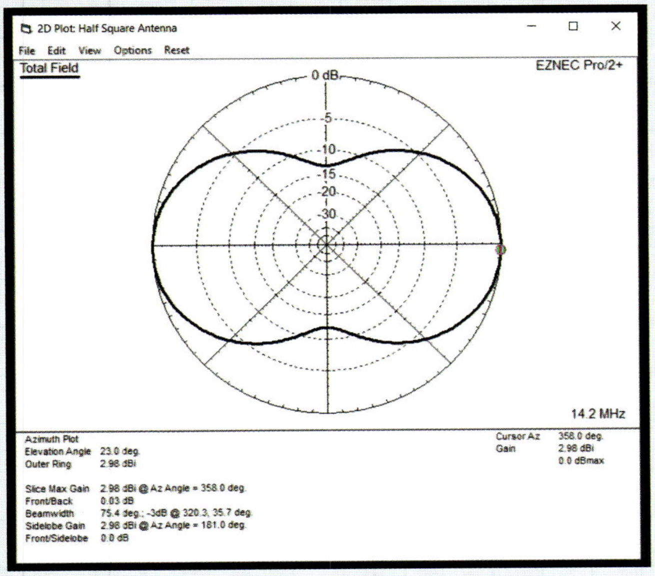

Typical Specifications

Low Takeoff Angle: It radiates at low elevation angles (typically 20° to 25°), making it highly effective for DX contacts compared to a standard horizontal dipole at the same height.

Bidirectional Gain: The antenna provides approximately 3 to 4.5 dB gain over a single vertical or dipole, firing broadside to the wire (perpendicular to the run of the horizontal section).

Quiet Reception: It is significantly quieter than a vertical antenna because it behaves as two phased verticals, helping to null out local noise from certain directions.

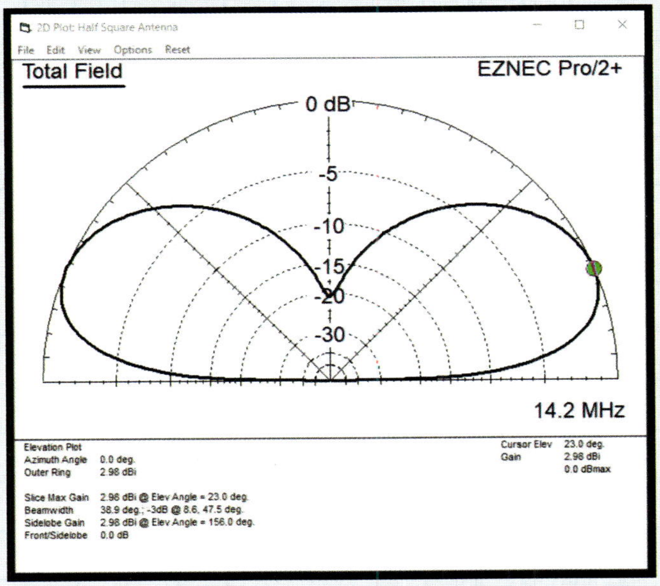

When building this antenna, consider that the feed point must be on the upper corner of the square, and the coax feedline must be as horizontal as possible coming away from the feed point. I've extended my coax away at a 45° angle with no issues, but if your coax is too close to the hanging quarter-wave element it will interfere with the phasing of the two vertical wires.

Using a 1:1 balun at the feed point makes this an excellent current-fed monoband antenna and provides a direct match to 50 ohm coaxial cable, requiring no matching network.

This antenna can also be constructed using one single wire, in the same shape and form, but end fed. This turns the antenna into a voltage-fed multiband antenna with the feed point located at the bottom corner of one vertical leg. Because this is a high-impedance point, it requires a 49:1 or 64:1 unun transformer. This configuration allows for multiband use on the fundamental and its harmonics (e.g., 40, 20, 15, and 10 meters). This form of the half square antenna completely changes its radiation pattern and cripples its DX capabilities compared to the true corner fed version of the half square.

If you have a couple poles, some wire, and a balun, give this antenna a try! Hopefully, you'll have one of those amazing DX moments like I did with the half square.

HALF SQUARE
ANTENNA FOR 20M

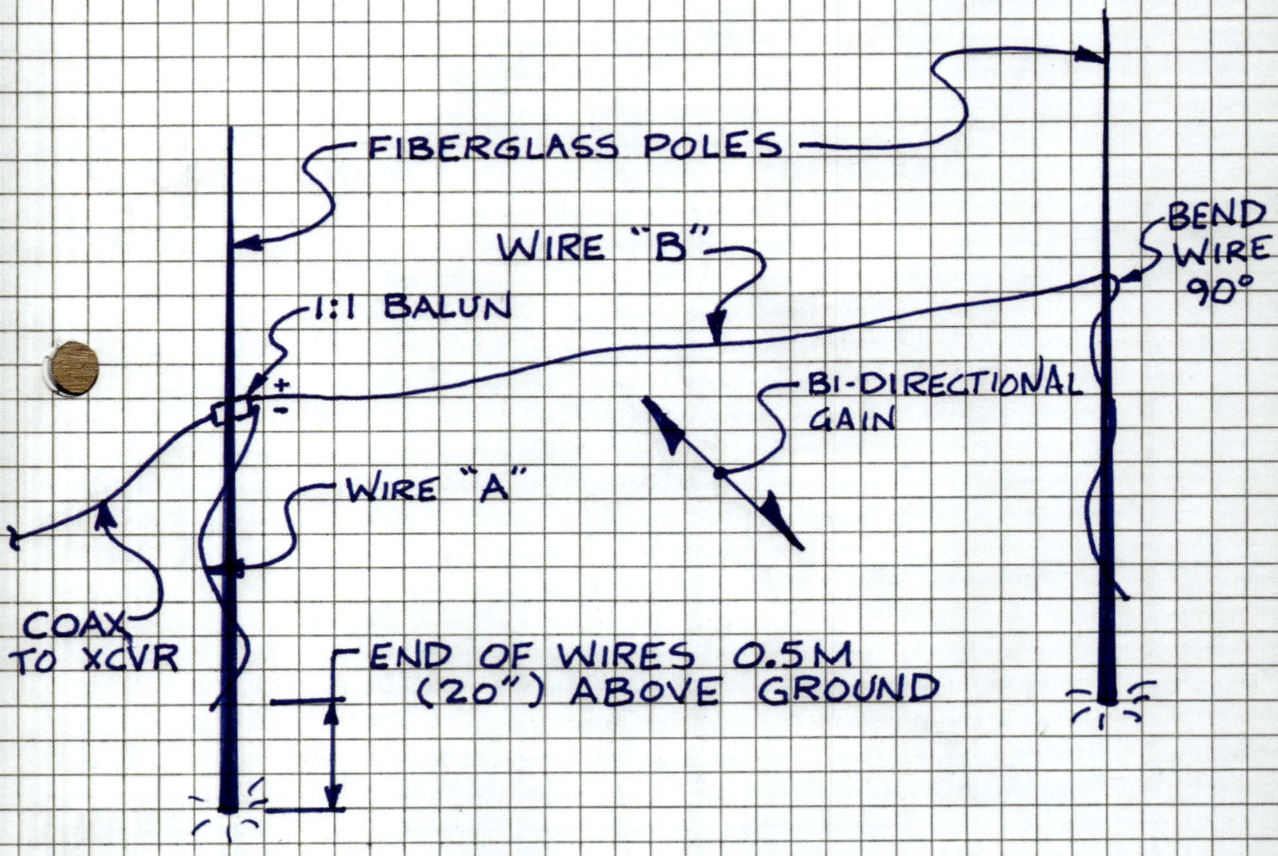

NOTES:

WIRE "A" LENGTH = 5.194 M (17'-0½")

WIRE "B" LENGTH = 15.581 M (51'-1 7/16")
10.387 M (34'-0 5/16") ACROSS
5.194 M (17'-0½") SIDE

WIRE IS COATED 22 GA

SALTY WALT K4OGO

Salty Walt's Portable Rigs

The zBitx

The **zBitx** is a 5 watt, 64-bit integrated SDR radio. It covers 80 to 10 meters, with CW, SSB, AM, FreeDV, FT8, SSTV, and a 480 × 320 pixel touch screen. It measures just 6.5 x 3 x 1.5″ and can run off two LiPo 18650 batteries. I have sat on the beach with this radio in the palm of my hand and have made some fun and amazing QRP contacts around the world. It has so many more features that I don't even use, including integrated FT8. If portable FT8 is your thing this might be the QRP rig for you. Made in India by HF Signals, this affordable little rig is one of my favorites for QRP operations.

The Bobtail Curtain

Taking the half square antenna from the previous chapter to the next level is quite easy. Simply add another pole and a piece of wire to make an antenna called the bobtail curtain. A third quarter-wave vertical radiating element gives the curtain even more gain. By mirroring the half square antenna's feed point configuration, we can create an extremely effective DX antenna.

A variation of the bobtail curtain is fed at the bottom of the center vertical wire. This feed point generates very high impedance and requires a matching unit and/or tuner. I built this antenna in its true "curtain" form, feeding the antenna at the top of the center vertical, which makes it a current-driven antenna. This configuration gives a great low takeoff angle for DX and is broadsided with good directional gain of 5 dB or more. I built this on the beach for 10 meters and made great DX contacts across the Atlantic Ocean from my QTH in Virginia Beach. One characteristic of this antenna that really stood out to me was how well it received and how quiet the noise level was. It was uncharacteristic of the normal noise of a vertical antenna.

Using a 1:1 balun at the feed point makes this an excellent current-fed monoband antenna and provides a direct match to 50 ohm coaxial cable, requiring no matching network.

Core Components

Wires: Two ¾ wavelength wires and one ¼ wavelength wire.
Balun: 1:1

Supports: Three 7 or 10 meter fiberglass poles.

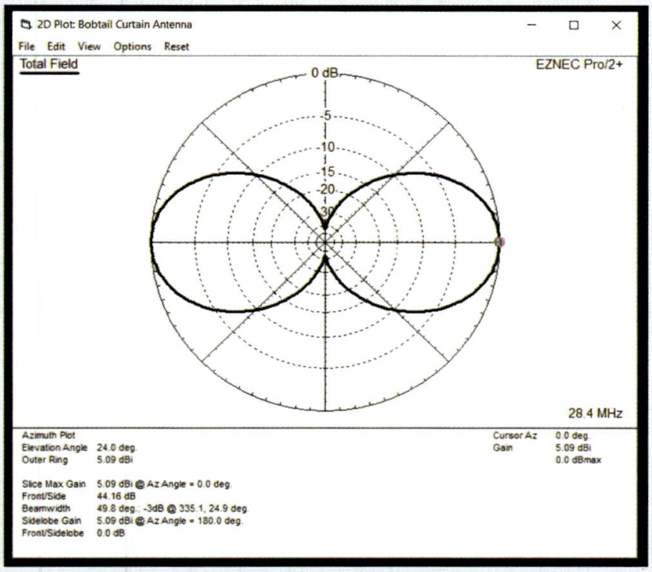

Typical Specifications

Low Takeoff Angle: The bobtail curtain radiates at low elevation angles (typically 20 – 25°), making it highly effective for DX contacts compared to a standard horizontal dipole at the same height.

Bidirectional Gain: The antenna provides 5+ dB gain over a single vertical or dipole, firing broadside to the wire (perpendicular to the run of the horizontal section).

Quiet Reception: It is significantly quieter than a vertical antenna, even more so than the half square, because it behaves as three phased verticals, helping to null out local noise off the ends of the wire.

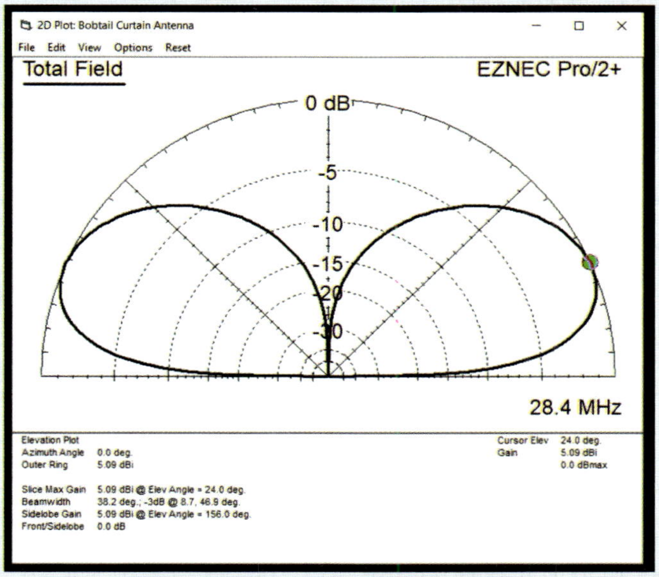

Add another pole if you have a third and use some wire and a balun. This antenna is another simple way to form a beam and direct your signal. It's an excellent DX antenna not only on the beach but anywhere you have the space to put it up. Give it a try!

BOBTAIL CURTAIN

VERTICAL WIRE BEAM ANTENNA

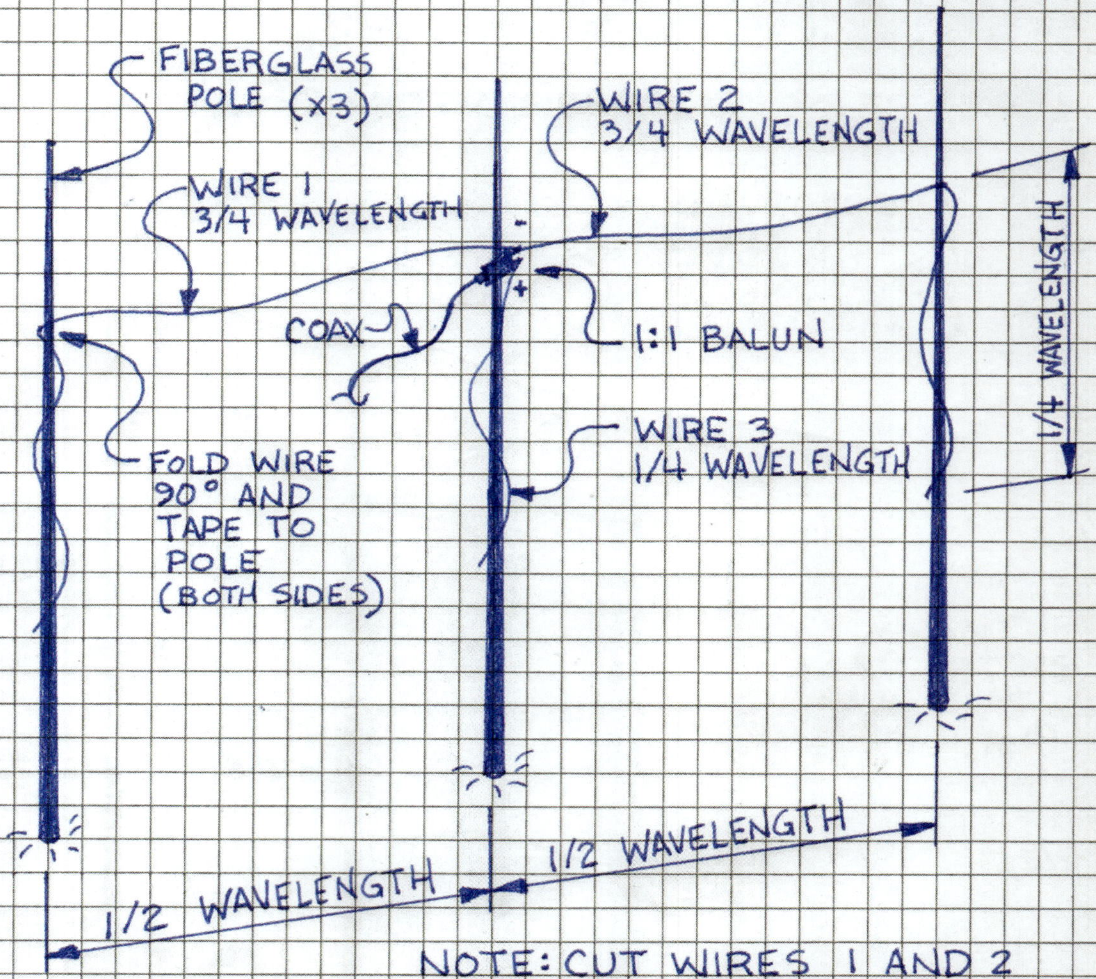

NOTE: CUT WIRES 1 AND 2 3/4 WAVELENGTH LONG AND FOLD TO FORM 1/2 WAVE HORIZONTAL AND 1/4 WAVE VERTICAL CONNECT BOTH TO (–) SIDE OF BALUN

SALTY WALT K4OGO

Salty Walt's Portable Rigs

The QMX+

The QRP Labs **QMX+** is a high-performance multimode transceiver that now officially supports Single Sideband (SSB) operation. While the QMX radios were originally launched as CW and digital-mode radios, they became full SSB transceivers following major firmware updates. This radio was the precursor to many of the USDX and other small QRP rig designs that piggybacked on its original concept. These radios are hand-built and not mass-produced like so many other cheap QRP rigs, and they're extremely affordable to boot. For pure QRP fun using SSB, I love this little rig!

Co-Phased Vertical Whips

In the previous chapters we have experimented with beamforming in its simplest form: taking wire and a single feed point to form a directional signal. To take beamforming to the next level we can use two quarter-wave vertical driven elements and form a beam by spacing them apart appropriately and feeding them in-phase or "co-phased" using a special harness. The design and build of this coax harness is the key to success when co-phasing two antennas, and fortunately there are several off-the-shelf harnesses available thanks to the popularity of co-phasing in the CB radio world. That's right, when you see a big tractor-trailer truck heading down the road with two antennas up high on its mirrors, those antennas are co-phased for the 11 meter band, and hams can use that same coax harness setup when building co-phased portable antennas, especially the 10 meter band.

Co-phasing two identical quarter-wave vertical elements spaced exactly one half of a wavelength apart creates a bidirectional antenna with higher gain that radiates broadside to the line of elements. The half square antenna operates on a similar principle, but using a phasing system—in its simplest form, a co-phasing harness—reduces the antenna system's losses and increases its gain.

The easiest way to build a co-phased antenna system is to purchase a pre-built harness, but before you go and take that shortcut, let's review what that harness is doing and see what it takes to build one.

Core Components

Antenna Elements: Two ¼ wavelength telescoping whips.
Spacing: ½ wavelength between whips.
Coaxial Cable: Use 75 ohm coax (like RG-59 or RG-6) for the phasing lines. Use 50-ohm coax for the main feed line to the radio.
Connectors: Use appropriate connectors (e.g., PL-259s and a female-to-female-to-female T-connector).
Radials: Each antenna must have the same number, arranged identically.
Tools: An ohm meter is crucial for checking for shorts, and a network analyzer can help verify the electrical length and SWR.

Typical Specifications

Gain: 3.27 dBi.
Sidelobe Gain: 3.27 dB
Elevation Angle: 27 °
Radiation pattern: Broadside

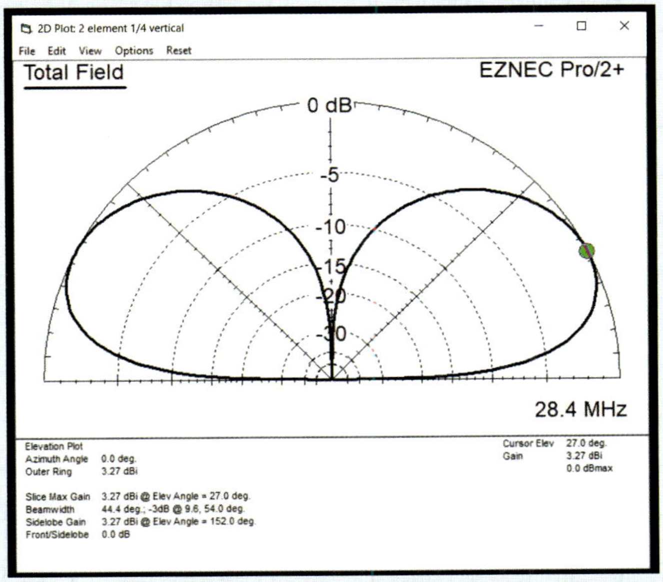

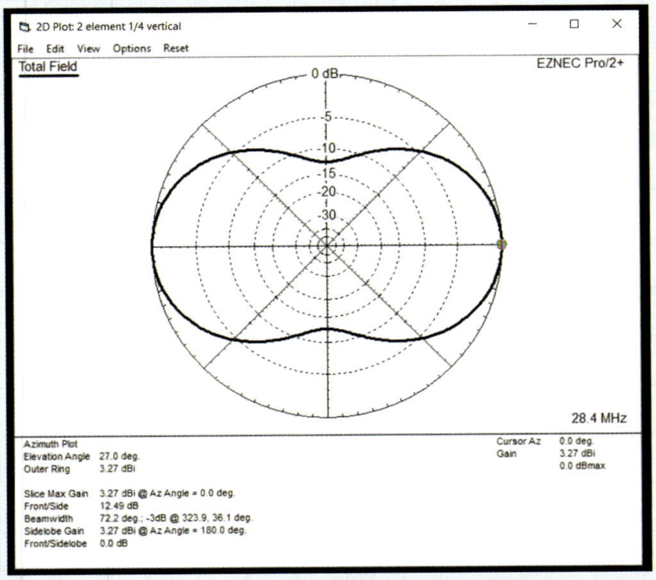

Design Principles and Components of a Co-Phase Harness

Impedance Matching: One of the harness' key functions is to transform the individual antenna impedances into a single 50 ohm load for the main feedline. To achieve this transformation, a two-antenna system uses quarter-wavelength (or odd multiples of a quarter-wavelength) sections of 75 ohm coax as impedance transformers. They convert the 50 ohm antenna impedance to 100 ohms at the connection point or "T-connector." The connector then sends a single 50 ohm feed line to the transceiver, resulting in a parallel impedance of 50 ohms.

Identical Electrical Lengths: The cables connecting each antenna to the common feed point must have exactly the same electrical length to maintain the correct phase relationship. The physical length calculation must account for the specific coaxial cable's velocity factor (VF) (e.g., around 0.66 for standard RG-59, but it can vary by manufacturer, so consulting the datasheet of the coax or using a network analyzer is recommended).

Construction Steps (Two Antennas)

For a basic two-antenna, in-phase array using the common T-connector method:

1. Determine the operating frequency: The design is frequency-dependent.
2. Calculate wavelength: Calculate the wavelength (λ) in free space using the formula: Wavelength (m) = Speed of Light (300,000,000 m/s) / Frequency (Hz).
3. Determine coax length: The connecting cables (L1 and L2) need to be an odd multiple of a quarter-wavelength.
   a. Calculate the quarter-wavelength in free space (λ/4).
   b. Multiply this length by the velocity factor (VF) of your specific 75 ohm coax to get the physical length.
4. Ensure both L1 and L2 are cut to this exact length.

OK, so there's the hard way to do it!

As I stated this can be done very easily on the 10 meter band using a prebuilt harness readily available from your local truck stop or CB radio supply. I have built a few portable co-phased setups this way, and have used two simple, identically tuned telescopic whip antennas spaced half a wavelength apart. The key to the setup was tuning each element individually to the exact same frequency using a normal piece of single 50 ohm coax on each vertical element and then attaching the co-phase coax harness. It takes a little time and effort, and you need to use an antenna analyzer or SWR meter to get it dialed in exactly, but the results of beaming your signal using this method are well worth the effort. We will discuss phasing and more advanced antenna phasing methods later in this book, but this method of antenna phasing provides an excellent entrée. Give it a try!

CO-PHASED VERTICAL ANTENNAS

FOR 10 METERS

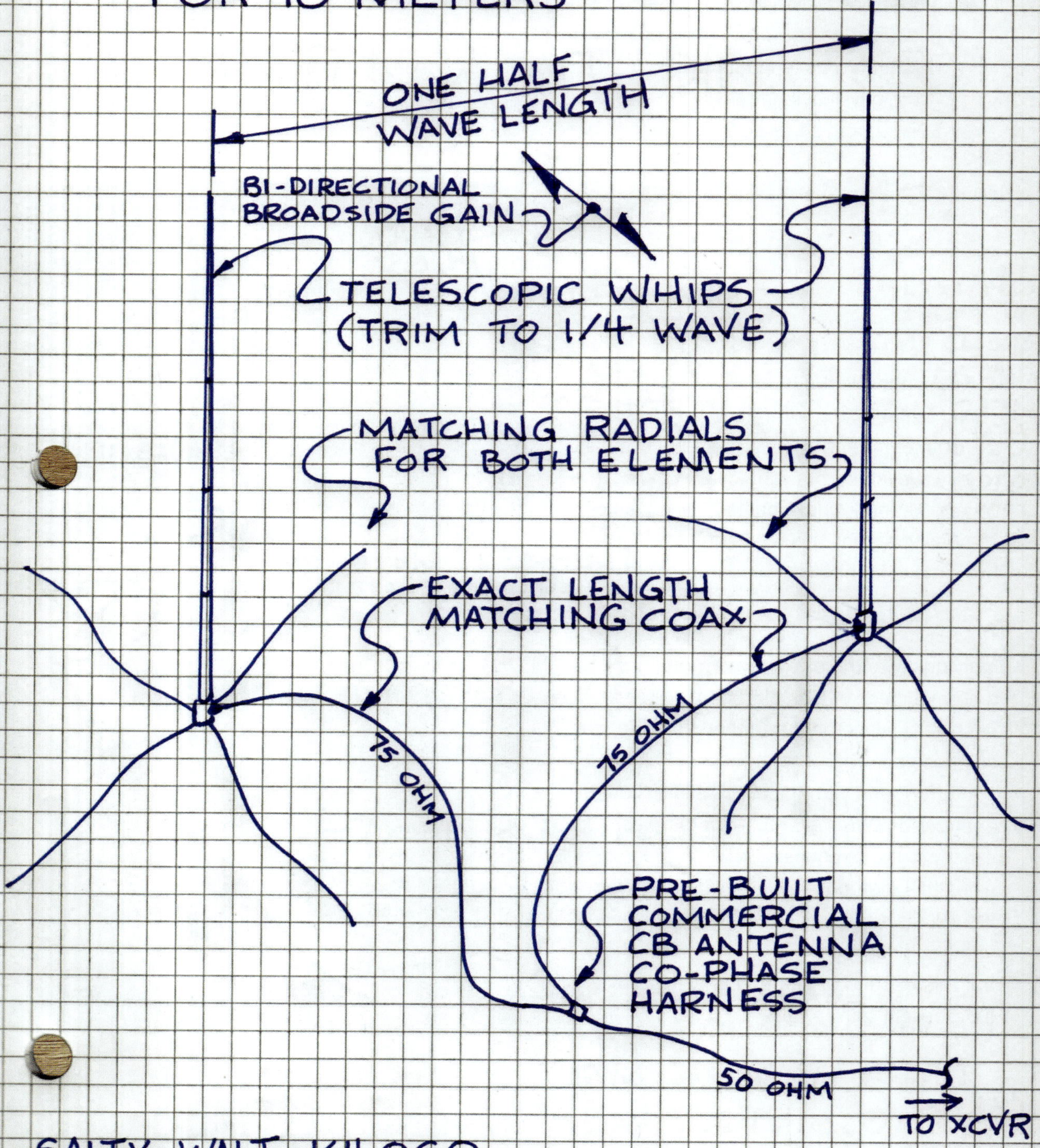

SALTY WALT K4OGO

Salty Walt's Portable Rigs

The uSDX

The **uSDX** (and its popular variants like the uSDR) is an ultra-compact, low-cost, open-source, QRP (low power), HF, ham radio transceiver. Originally developed as a modification for the QCX CW kit, it has evolved into an inexpensive SDR radio capable of all-mode operation. Unfortunately, this radio has been cloned and produced cheaply, and sometimes finding a solid quality **uSDX** rig can be a challenge. There is a lot of junk out there. I've found that purchasing from a source with the ability to return it quickly is the best option. I have really enjoyed using this little radio as a hand-held on the beach as they have their own internal battery. My two favorite **uSDX** rigs are what I call "The Red Brick" which is the popular metal case version and the "Red Corner" plastic cased version. I have had some great QRP fun with these little rigs.

The Corner-Fed Cubical Quad

The cubical quad antenna is an improvement on the Yagi antenna whose invention was driven by necessity. The antenna was created for the shortwave missionary radio station HCJB in Quito, Ecuador, which was increasing its transmitter power to 10,000 watts. The high altitude and specific atmospheric conditions in the Andes caused electrical arcing on the tips of their four-element Yagi antenna, damaging the equipment. HCJB engineer Clarence Moore, W9LZX, solved the problem in the early 1940s by using a full-wave loop principle. Moore's design insight was to pull open a folded dipole, which is usually rectangular in shape with its short sides significantly less than half a wavelength long, into a full-wave square loop. This closed-loop design has a lower voltage at its extremities than a Yagi, which helps it solve the corona discharge problem the HCJB Yagis were experiencing. To give the antenna directional capabilities like a Yagi, the cubical quad uses a driven element and a slightly larger parasitic element (reflector) placed behind it. Slightly shorter elements in front of the driven element increase the antenna's gain. It's simply a beam antenna made out of square or "cubical" elements.

Core Components

Antenna Elements: Precut cubical quad reflector and driven element.
Spacing: Approximately 1/4 wavelength between elements. The driven element should be 2.035 meters above ground, and the reflector should be 2 meters.
Feed point: Feed the driven element in a bottom corner with a 4:1 balun using 50 ohm coax.
Poles: Four 7 meter fiberglass poles.

Typical Specifications

Gain: 8.09 dBi.
Sidelobe Gain: –3.24 dB
Elevation Angle: 24°
Directionality: Perpendicular to the plane of the squares.

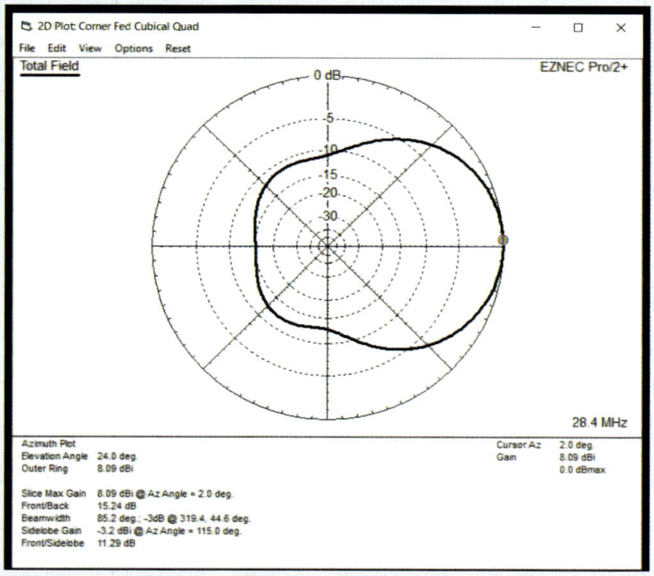

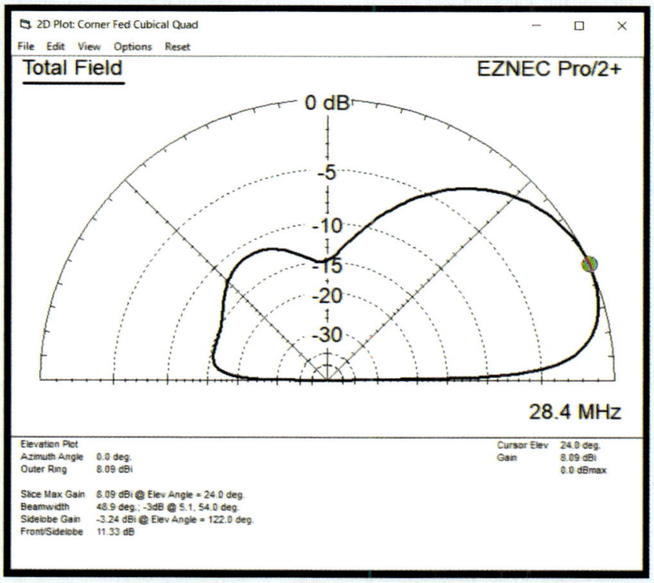

Some of the key advantages of the cubical quad are:

Higher Gain: A well-optimized cubical quad generally provides about 2 dB more gain than a Yagi with the same boom length and number of elements.

Lower Noise Reception: Due to its closed-loop nature and reduced susceptibility to precipitation, static, and common mode feedline currents, it tends to be quieter on receive, which improves the signal-to-noise ratio.

Wider Bandwidth: The quad's inherently low "Q" factor provides a wider bandwidth, meaning it can operate efficiently across a greater range of frequencies without extensive re-tuning.

Lower Radiation Angle: It offers a lower angle of radiation compared to a single Yagi, which is advantageous for long-distance (DX) communication.

I found a way to build a cubical quad in its simplest form with just a driven element and a reflector using two wires and four poles. The issue with this design was the feed point, and the easiest way to overcome this was to shape the wires into a simple square and not the upright diamond shape that is typical for cubical quads. With this square shape I could feed the driven element in the lower corner by taping a 4:1 balun to a pole. To construct this antenna, I simply hammered four pieces of PVC pipe into the ground, making sure to properly space them, and removed the bottom caps of my four DX Commander 7 meter poles. Then, I extended the poles, laid them out on the ground, and taped my premeasured cubical wires to the poles at the appropriate height. Finally, I picked up two of the poles at a time and slid each one over a PVC base and repeated the process until all four poles were up. I did this by myself during a portable activation.

My design worked great for the 10 meter band and also had a workable SWR on the 11 meter band, so CB hobbyists could use it as well. The antenna is a large antenna and more suitable for portable builds in the higher bands as shown in my sketch and build, but for those that want to experiment on lower HF bands it can be done with larger support poles and bigger quad boxes. If you're feeling adventurous, search online for a cubical quad calculator and give it a try.

Modelling for my antenna build showed an 8 dB directional gain. Using it proved to me that the model was probably accurate. If you want to play with an interesting antenna with incredible directional gain, grab four fiberglass type poles, cut two wires and go make some great DX contacts.

CORNER FED CUBICAL QUAD

ANTENNA FOR 10 METERS

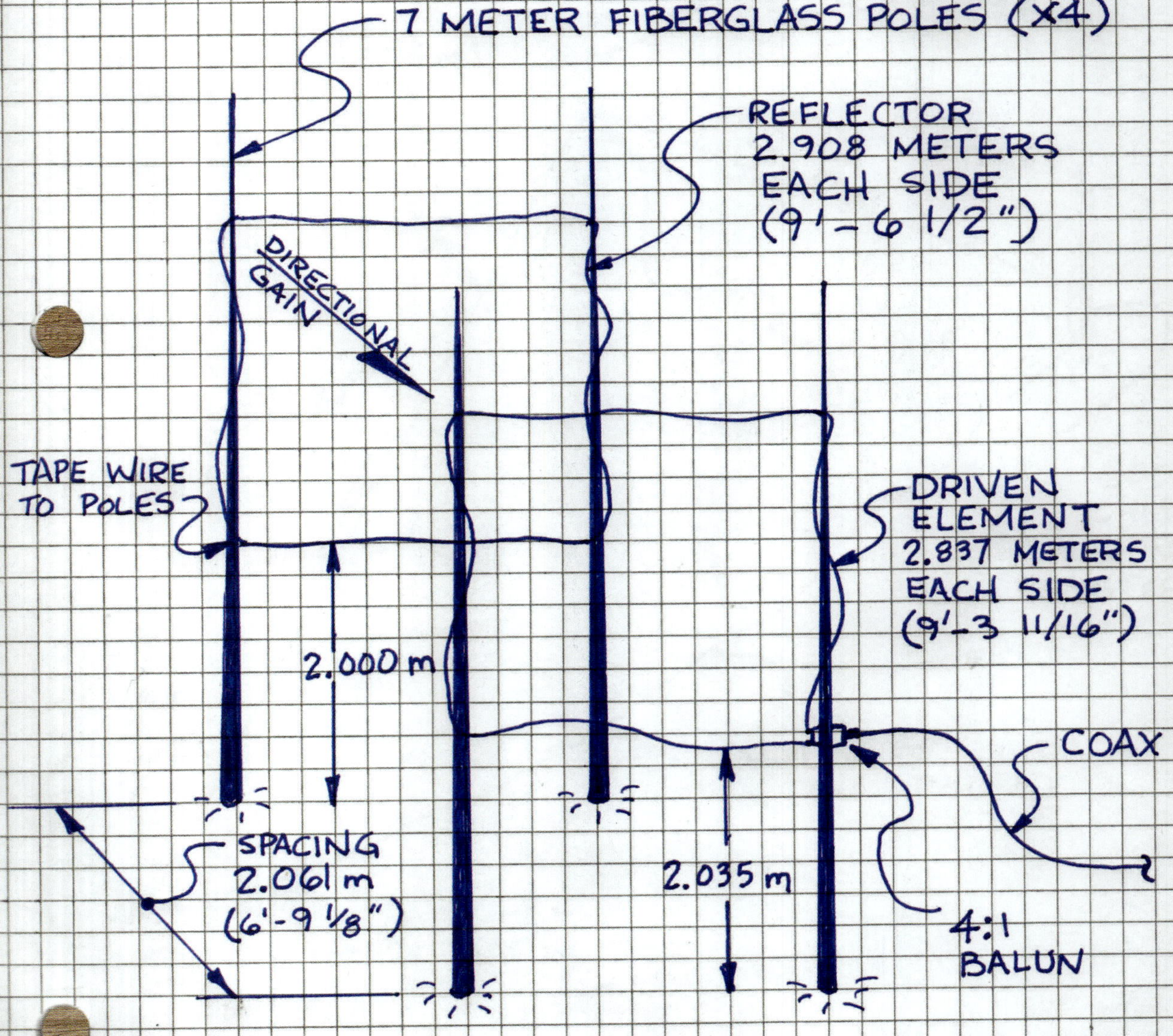

SALTY WALT K4OGO

Salty Walt's Portable Rigs

The FT-450D

The transceiver that turned me into a Yaesu fan is the **FT-450D**. I was looking for an affordable used 100 watt rig with an internal tuner (ATU) and found a bargain on this radio. The **FT-450D** improved on the original FT-450 by adding the tuner and some other upgrades. It was introduced in 2011 and continued production for several years. On the used market you can still find them in good shape. Its size is perfect for portable use and it's a simple radio to use, making it great for a newer ham. I love my Yaesu **FT-450D**.

The Portable Rhombic

When it comes to portable antennas, nobody thinks of the rhombic because it's so large. But should they? Just what is a rhombic antenna? The rhombic antenna, designed in 1931 by Edmond Bruce and Harald Friis at Bell Telephone Laboratories, is a broadband, directional, diamond-shaped antenna crucial for early long-distance HF (shortwave) communication. Renowned for its high gain, it dominated point-to-point, military, and international broadcasting before WWII. It consists of four equal-length wires forming a rhombus shape, usually supported by four masts, which can be bidirectional or, if terminated with a resistor at one end, unidirectional. This antenna was so popular in the early days of radio that many radio societies around the world including the ARRL use the rhombus shape in their logos. The design saw heavy use during World War II as a receiving antenna. Following the war, it was largely replaced by more efficient, compact designs like log-periodic antennas and curtain arrays.

Core Components

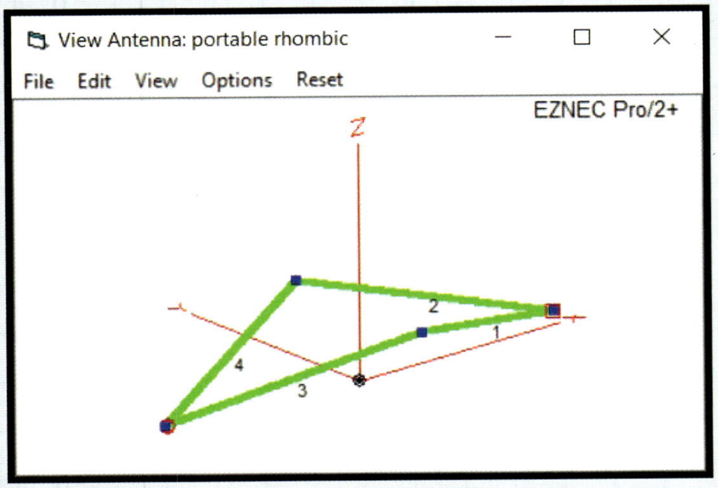

Size: They are physically large, typically requiring multiple wavelengths for optimal performance (e.g., 100+ feet per leg).
Termination: A non-inductive resistor is used at the end opposite the feed point to create a unidirectional pattern, though power is lost in this resistor.
Impedance: Features high characteristic impedance, usually around 450 – 600 ohms, requiring a 12:1 or similar balun for use with 50 ohm coaxial cables.

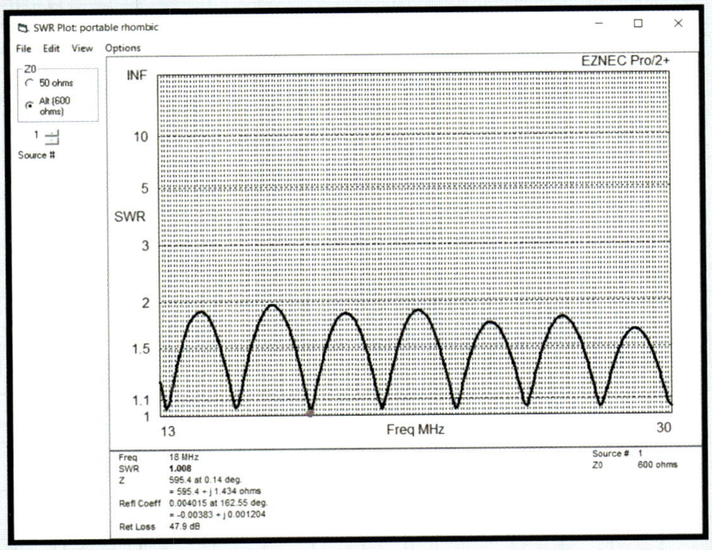

Typical Specifications

High Gain and Directivity: Provides high gain with low take-off angles, ideal for long-distance communication (DX).
Broadband Performance: Capable of operating over a wide range of frequencies.
Design and Construction: Dimensions are not critical, making them popular for high-power, permanent base stations.

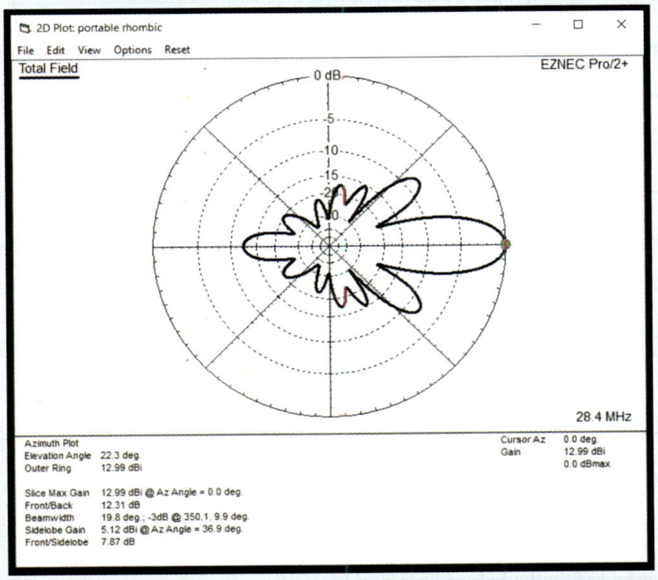

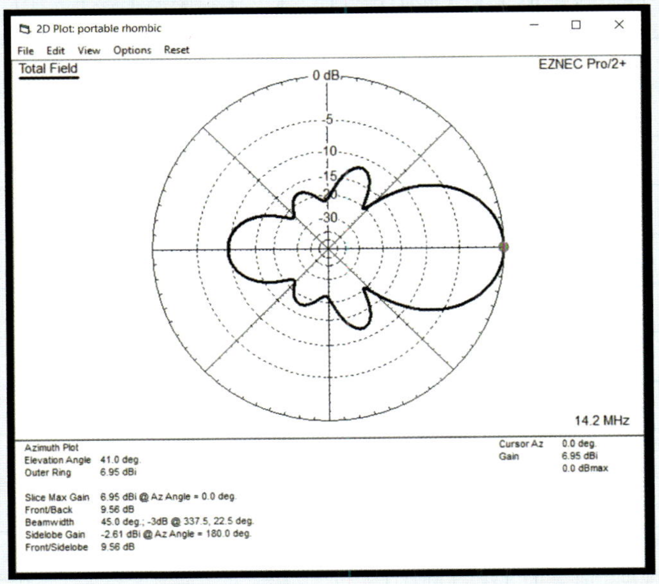

(continued next page)

Typical Specifications

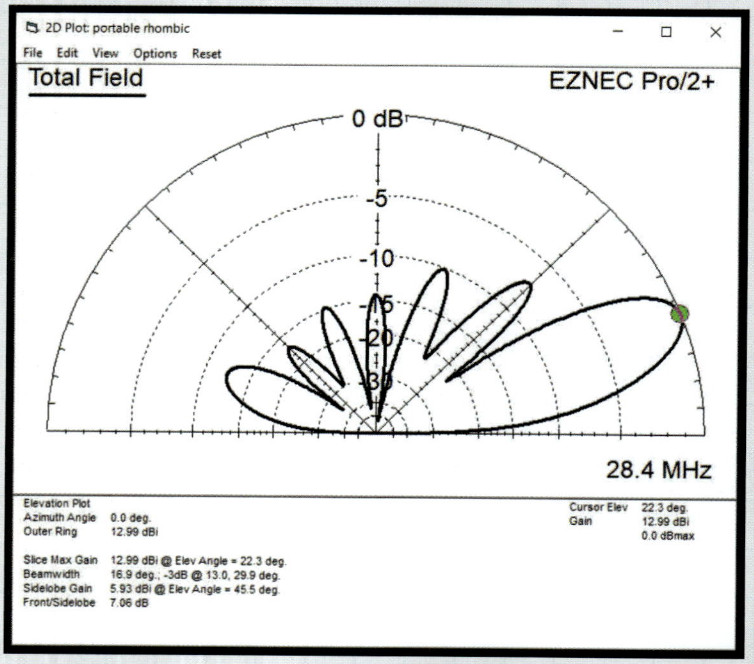

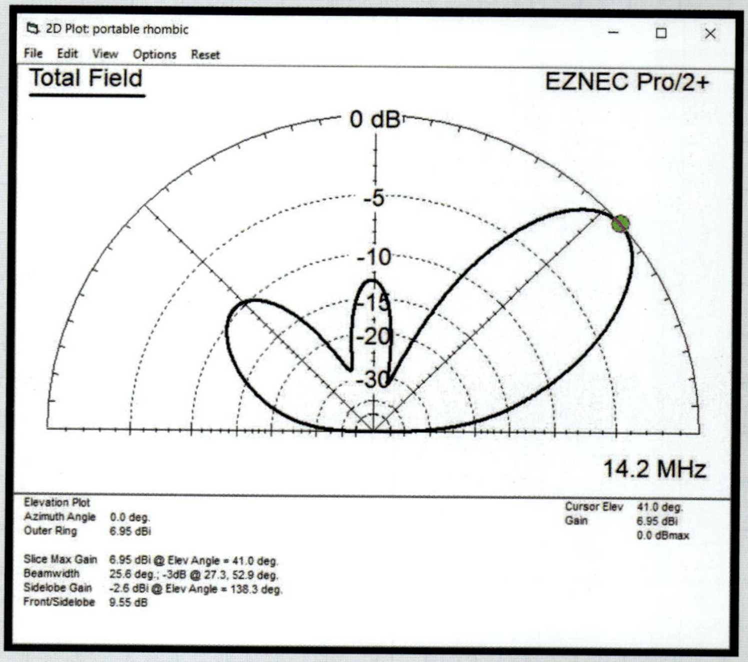

The antenna is often considered the "king of wire antennas" for its effectiveness despite its significant space requirements, often with legs several wavelengths long.

Taking its characteristics into consideration, I set out to build a "portable" version of the rhombic. My design only used two large 10 meter poles at the antenna's mid-point, meaning the wires slope down toward each end. Because of the weight of the wires, the poles bent over, but otherwise held up fine to support the elements.

This configuration is not as efficient at a true, flat, horizontal rhombic antenna, but this setup made the antenna easier to build for portable use. As this antenna also is more effective the longer the wire legs are, I found that the minimum length I could use in a portable setting was 100 feet per leg which is approximately 1.5 wavelengths for the 20 meter band and 3 wavelengths for the 10 meter band. These lengths for each leg made for great DX characteristics on every band from 20 through 10. It worked on the lower 80 through 30 meter bands but had NVIS characteristics on these bands due to the shorter wire legs and the sloping ends.

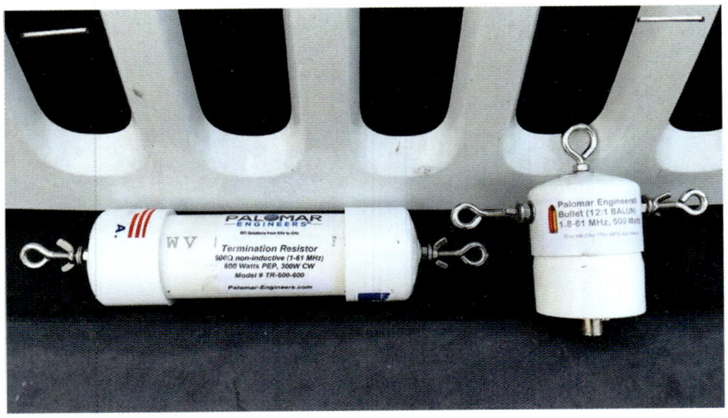

The heart of this antenna, as built, is the 12:1 balun and the 600 ohm terminating resistor, which I purchased from Palomar Engineering in California. Feeding this antenna with 50 ohm coax and balancing the antenna with the resistor and balun gave the antenna a workable SWR of under 2:1 on every band from 80 to 10 meters without requiring a tuner.

This antenna took a lot of effort to build as a portable setup, but it works extremely well and is a DX beast. It's probably best to set up this antenna when you've got a weekend or even longer to operate rather than to try to set it up for a quick POTA activation or something similar. If you have a large field or some other space to build it, you will definitely be glad you tried a rhombic.

PORTABLE RHOMBIC ANTENNA

SALTY WALT K4OGO

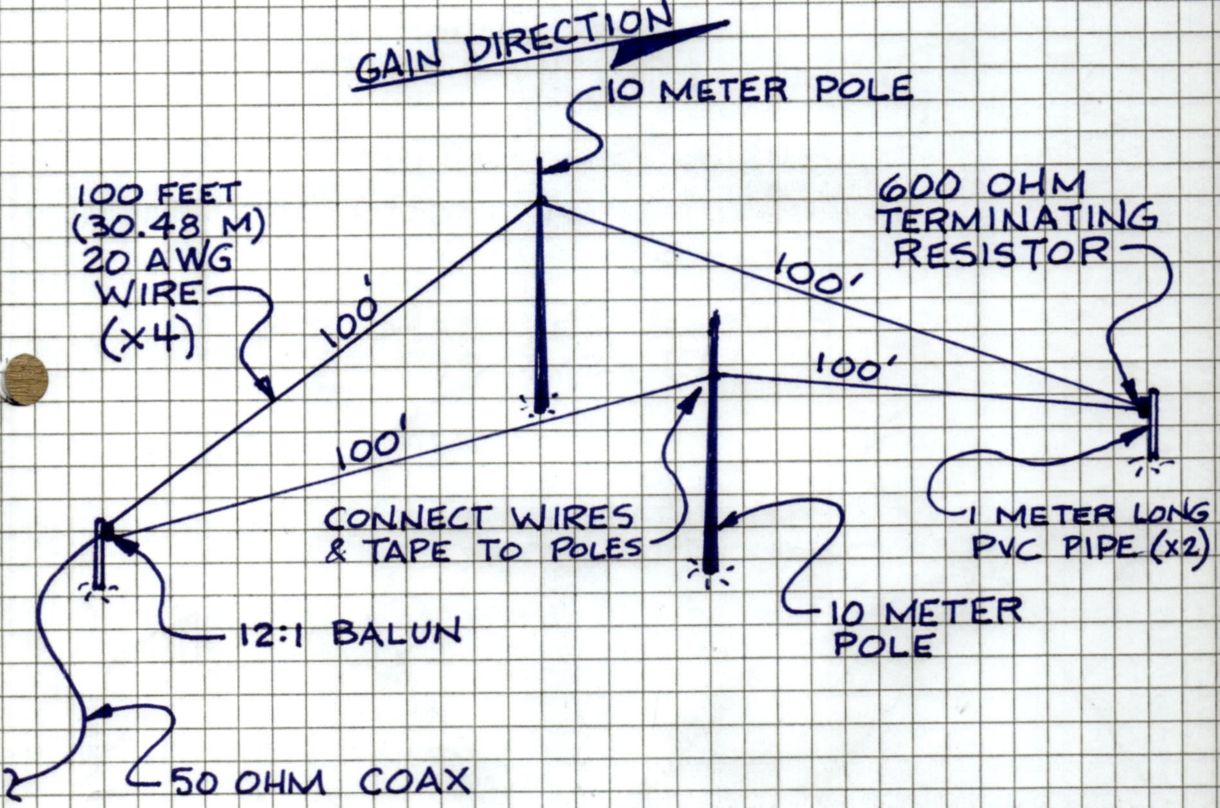

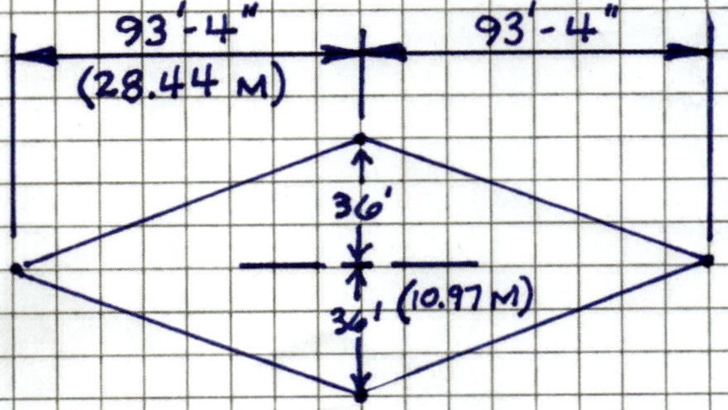

TOP VIEW DIMESIONS

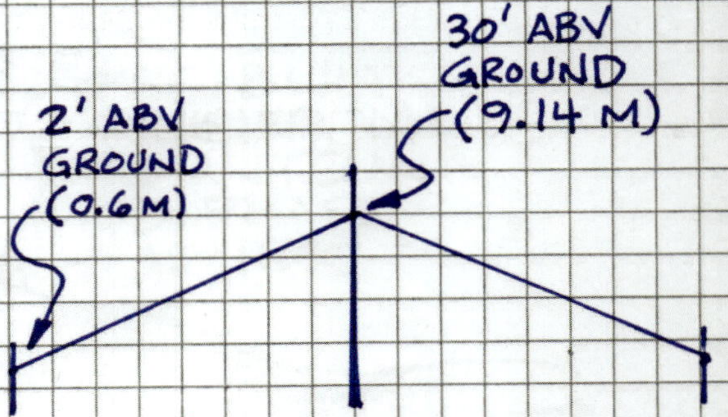

SIDE VIEW DIMENSIONS

Salty Walt's Portable Rigs

The FX-4CR

The **FX-4CR** is a compact, 20 watt software defined radio (SDR) transceiver covering the 80 through 6 meter bands. It operates USB, LSB, CW, AM, and FM, and the power is adjustable from 1 to 20 watts, making it a great QRP option when looking for something small and portable. It has a 2.0 inch TFT color screen with a spectrum display and 48 kHz waterfall plot, as well as built-in Bluetooth® and a USB-C port for digital modes and PC connectivity. I operated with it daily for two months while working in Europe as SP/K4OGO and received hundreds of great signal reports. No regrets. It's an excellent little rig with incredible audio for its small size.

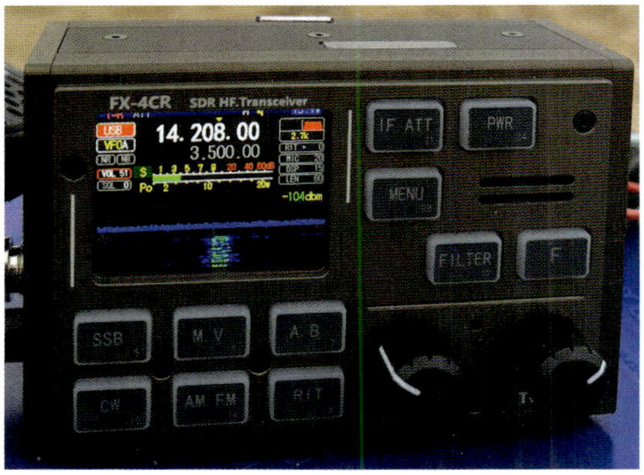

The Terminated End-Fed V

The terminated end-fed V (TEFV) is a rugged, broadband, nonresonant HF antenna popular in military radio for its simple, portable, tuner-free operation from 1.8 to 30 MHz. It utilizes a 9:1 or 12:1 balun at the feed point and a terminating resistor at the other end to minimize standing waves. It provides low-noise, is DX-friendly, and has NVIS capabilities as well. It can be found in nearly all unclassified US military HF antenna field books. This is a great military HF antenna to build and deploy as a portable ham radio antenna. To build this antenna true to military specifications requires one of the aforementioned baluns and a 450 to 800 ohm terminating resistor. The terminating resistor absorbs power and prevents high standing waves. Like many other antennas this one has been called by other names like the "half rhombic" and "inverted V with a counterpoise."

Core Components

Balun: 9:1 or 12:1
Resistor: 600 or 800 ohm
Wire length: Not critical. Shorter lengths decrease efficiency.
Counterpoise: A wire running along the ground or slightly above completes the circuit back to the transformer, improving SWR across all bands.
Mast: One 10 meter or longer.
Stakes: Two one meter lengths of 3/4 inch PVC.

Typical Specifications

Deployment: Designed for quick, tactical, or field deployment using trees or masts.
Polarization: The radiation pattern is a mix of vertical and horizontal, resulting in elliptical polarization.
Traveling Wave Design: Unlike resonant antennas that create standing waves, a TEFV uses a terminating resistor to absorb power that would otherwise reflect, creating a continuous current distribution.
High Gain and Directivity: Provides high gain with low take-off angles on the higher HF bands, ideal for long-distance communication (DX) and is very directional on the higher HF bands.
Near Vertical Incidence Skywave: Provides excellent NVIS capabilities on the lower HF bands.
Broadband Performance: The antenna operates efficiently over a very wide bandwidth without needing an antenna tuner. It typically provides a SWR of less than 2:1, or below 3:1 across all bands and can be used for multiband operation from 10 to 160 meters.

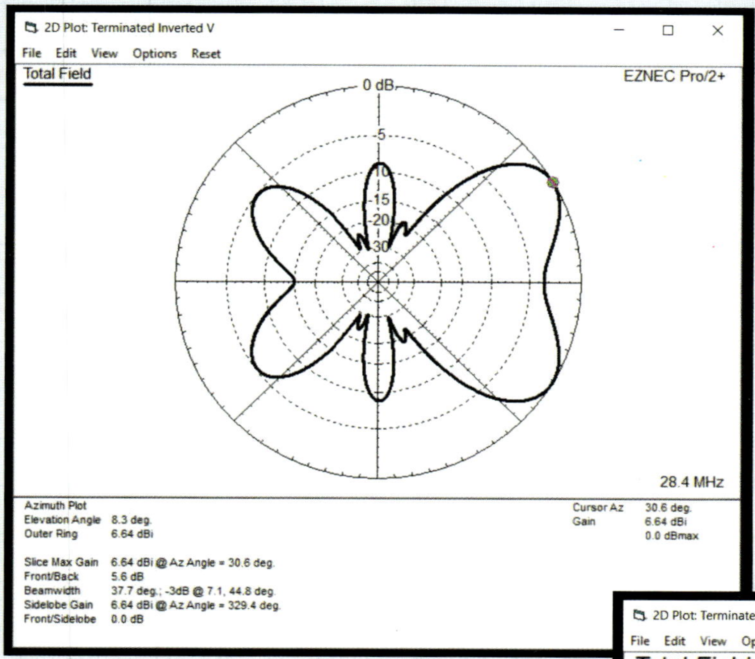

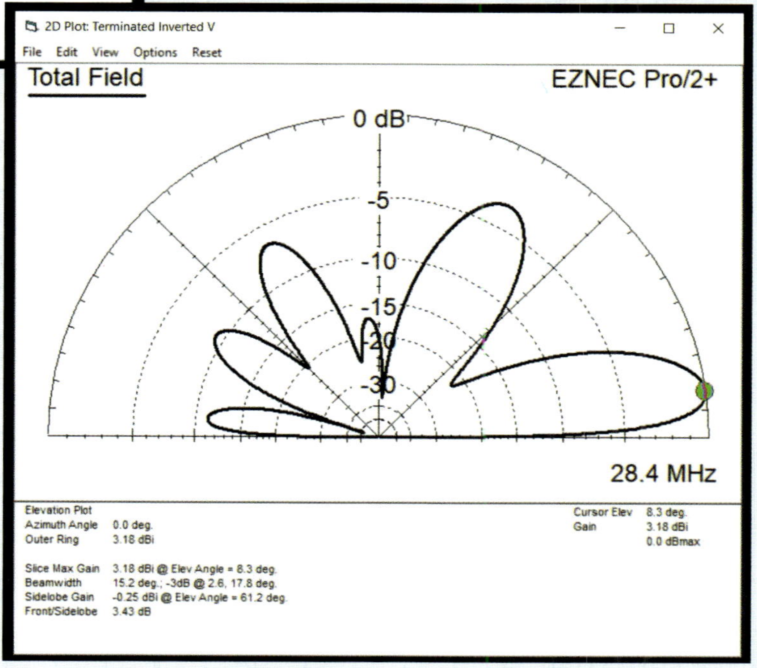

This antenna shares many characteristics with the rhombic antenna, but it is easier to deploy. It's basically half of a rhombic turned vertical. The counterpoise acts like a mirror to complete the other half of the antenna.

After reading about this antenna in every military manual I could get my hands on, I set out to build a version that was easy to set up with a single pole and a couple of smaller stakes or short pieces of PVC pipe. The pipe at one end of the antenna supports the balun and the pipe at the other end supports the terminating resistor. I used my DX Commander 10 meter Expedition pole and two short pieces of 3/4 inch PVC pipe about one meter long each that I drove into the ground with a rubber mallet.

The heart of the antenna I built is the 12:1 balun and the 600 ohm terminating resistor that I purchased from Palomar Engineering in California. Feeding this antenna with 50 ohm coax and balancing the antenna with the resistor and balun gave the antenna great SWR on every band. Just like a rhombic, the antenna didn't require a tuner.

If you're planning a portable setup, this antenna requires much less effort to build than the rhombic does, and it's a DX beast on the higher bands. I built mine during the ARRL 10 Meter Contest and logged worldwide DX stations with ease using 100 watts. Surprisingly, I also contacted nearby stations on 40 meters with solid S9 reports. That was definitely NVIS at its best. If you have the space to put up the TEFV, it is well worth experimenting with and building. This antenna provides a whole lot of juice for the squeeze.

TEFV ANTENNA

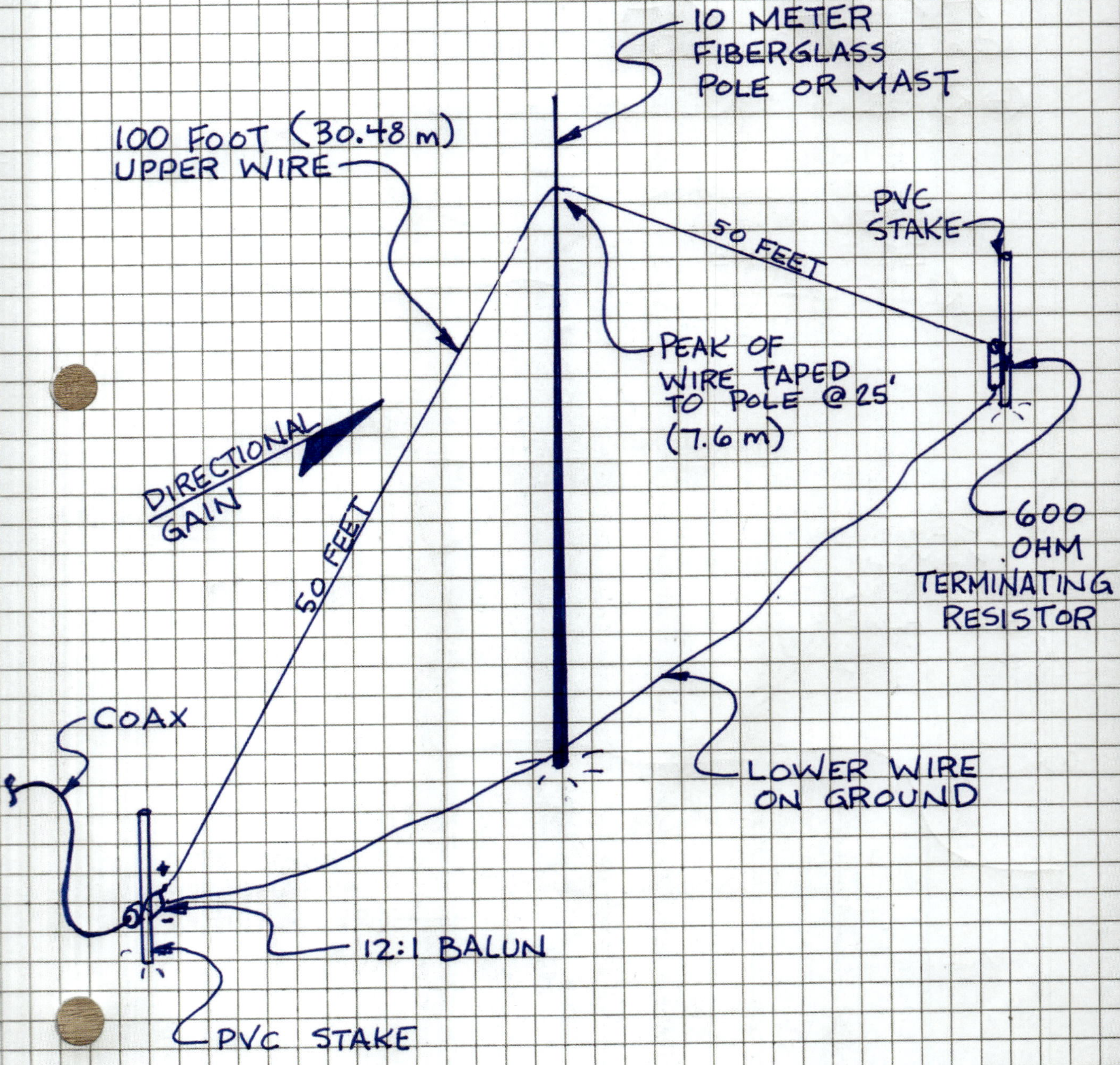

SALTY WALT K4OGO

Salty Walt's Portable Rigs

The PMR-171

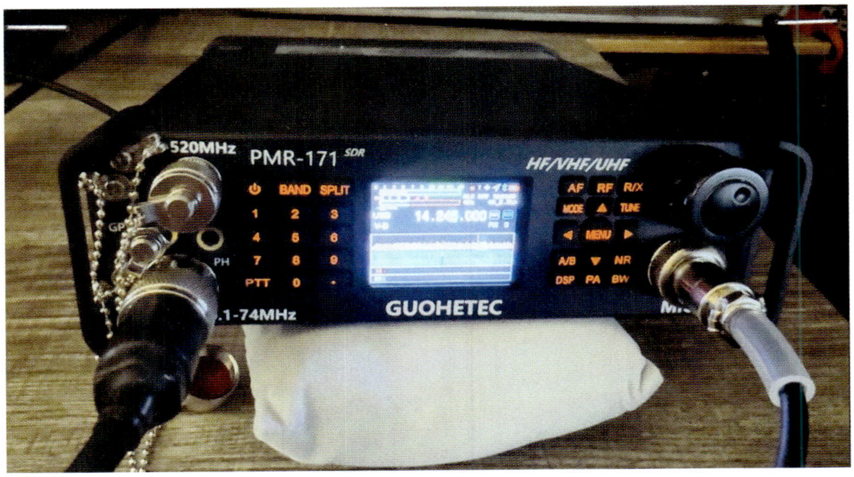

The Guohetec **PMR-171** transceiver is a compact, 20-watt software defined radio (SDR). It covers 160 – 10 meters, 6 meters, 2 meters, and 70 centimeters and operates USB, LSB, CW, AM, RTTY, FT8, DMR, and FM with adjustable power from 1 to 20 watts. It has a 2.0 inch TFT color screen with a spectrum display and a 48 kHz waterfall plot as well as built-in Bluetooth. An internal antenna tuner (ATU) makes experimenting with nonresonant antennas enjoyable. A military manpack-style radio, the **PMR-171** includes a detachable,14 watt battery, making it a great all-in-one portable transceiver.

The V-Beam

The V-Beam antenna is another military inspired HF field antenna. Like the TEFV, this antenna can also be found in several configurations in unclassified US military antenna manuals. It is a high-gain, directional, long-wire antenna consisting of two wires arranged in a V shape. Typically, it's fed with a balanced line or balun. Often used in HF radio for its high gain and narrow-beamwidth (with legs four wavelengths long, it's gain is comparable to a three-element Yagi), it's ideal for fixed-direction, long-distance (DX) communication on the higher HF bands (20 – 10 meters). Unterminated, this antenna is bi-directional, but if you add terminating resistors, you can point it in one direction.

Core Components

Balun: 9:1 or 12:1
Resistor: 600 or 800 ohm
Wire length: Not critical. Shorter lengths decrease efficiency.
Counterpoise: A wire running along the ground or slightly above completes the circuit back to the transformer, improving SWR across all bands.
Mast: One 10 meters or longer.
Stakes: Two one meter lengths of 3/4 inch PVC.

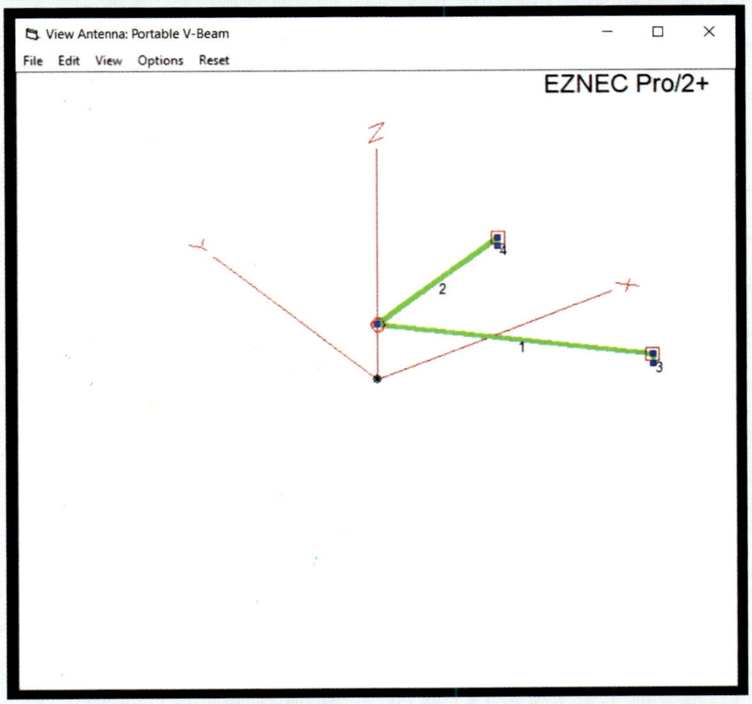

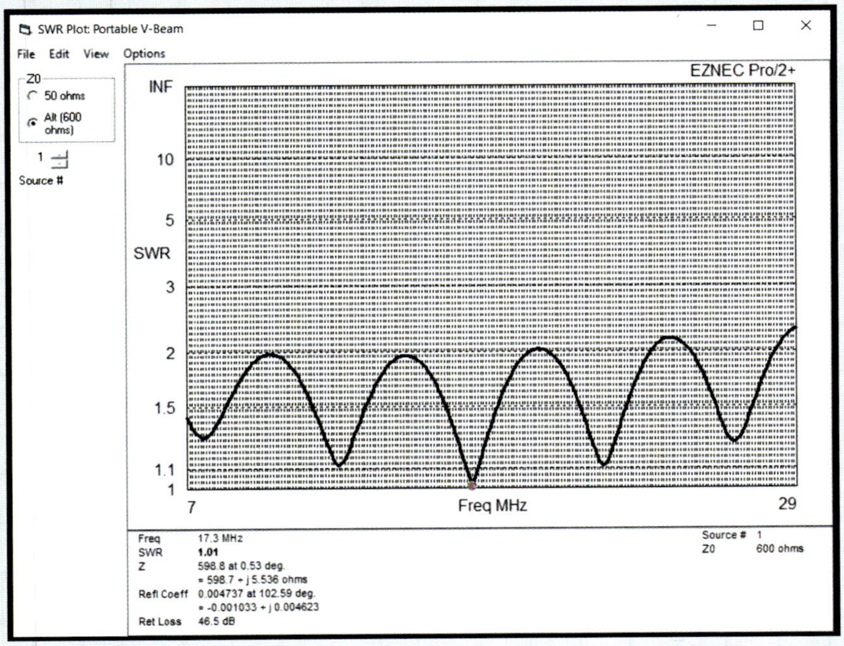

Typical Specifications

Pros: High gain, excellent for fixed-direction DX, broadband, and relatively simple to construct.
Cons: Requires significant space, narrow beamwidth, and requires a tuner for multiband use.

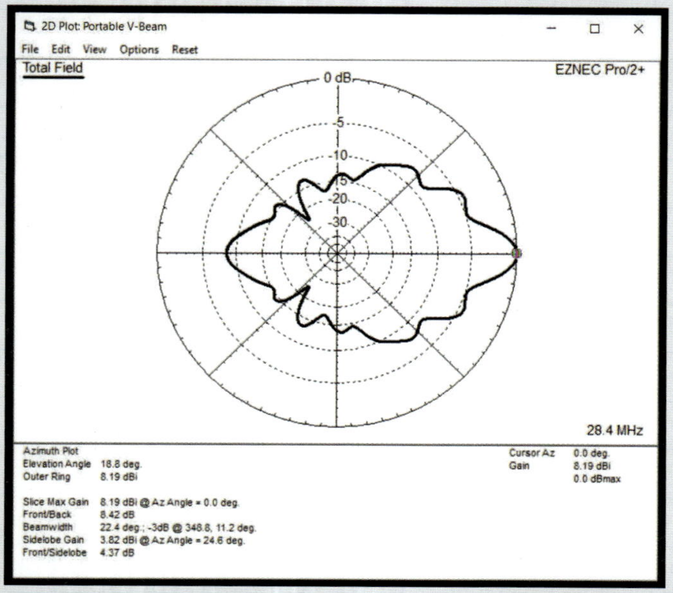
2D Plot: Portable V-Beam
File Edit View Options Reset
Total Field
EZNEC Pro/2+
0 dB
28.4 MHz
Azimuth Plot
Elevation Angle 18.8 deg.
Outer Ring 8.19 dBi
Slice Max Gain 8.19 dBi @ Az Angle = 0.0 deg.
Front/Back 8.42 dB
Beamwidth 22.4 deg.; -3dB @ 348.8, 11.2 deg.
Sidelobe Gain 3.82 dBi @ Az Angle = 24.6 deg.
Front/Sidelobe 4.37 dB
Cursor Az 0.0 deg.
Gain 8.19 dBi
0.0 dBmax

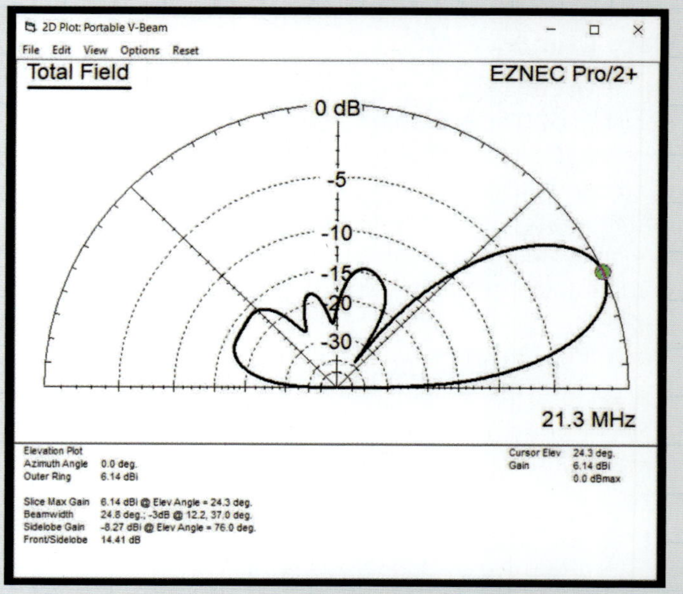
2D Plot: Portable V-Beam
File Edit View Options Reset
Total Field
EZNEC Pro/2+
0 dB
21.3 MHz
Elevation Plot
Azimuth Angle 0.0 deg.
Outer Ring 6.14 dBi
Slice Max Gain 6.14 dBi @ Elev Angle = 24.3 deg.
Beamwidth 24.8 deg.; -3dB @ 12.2, 37.0 deg.
Sidelobe Gain -8.27 dBi @ Elev Angle = 76.0 deg.
Front/Sidelobe 14.41 dB
Cursor Elev 24.3 deg.
Gain 6.14 dBi
0.0 dBmax

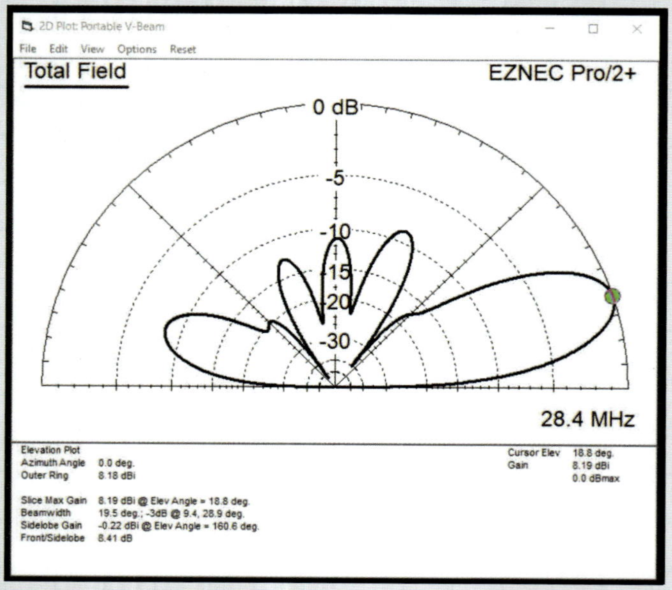
2D Plot: Portable V-Beam
File Edit View Options Reset
Total Field
EZNEC Pro/2+
0 dB
28.4 MHz
Elevation Plot
Azimuth Angle 0.0 deg.
Outer Ring 8.18 dBi
Slice Max Gain 8.19 dBi @ Elev Angle = 18.8 deg.
Beamwidth 19.5 deg.; -3dB @ 9.4, 28.9 deg.
Sidelobe Gain -0.22 dBi @ Elev Angle = 160.6 deg.
Front/Sidelobe 8.41 dB
Cursor Elev 18.8 deg.
Gain 8.19 dBi
0.0 dBmax

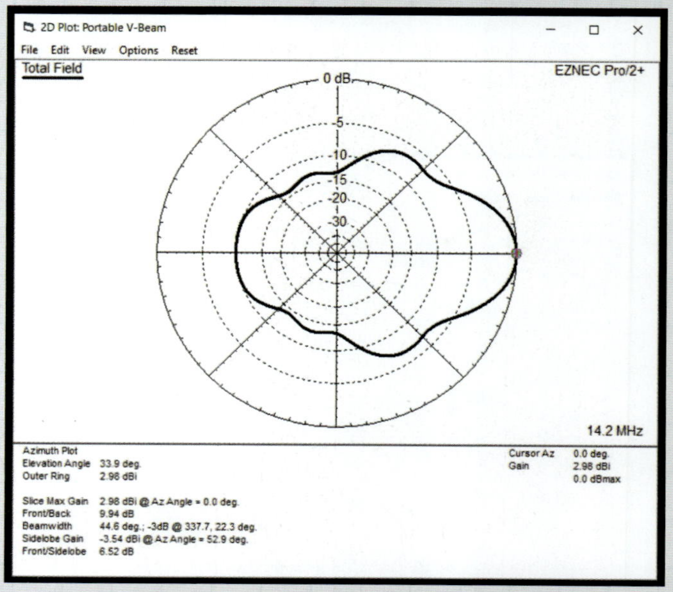
2D Plot: Portable V-Beam
File Edit View Options Reset
Total Field
EZNEC Pro/2+
0 dB
14.2 MHz
Azimuth Plot
Elevation Angle 33.9 deg.
Outer Ring 2.98 dBi
Slice Max Gain 2.98 dBi @ Az Angle = 0.0 deg.
Front/Back 9.94 dB
Beamwidth 44.6 deg.; -3dB @ 337.7, 22.3 deg.
Sidelobe Gain -3.54 dBi @ Az Angle = 52.9 deg.
Front/Sidelobe 6.52 dB
Cursor Az 0.0 deg.
Gain 2.98 dBi
0.0 dBmax

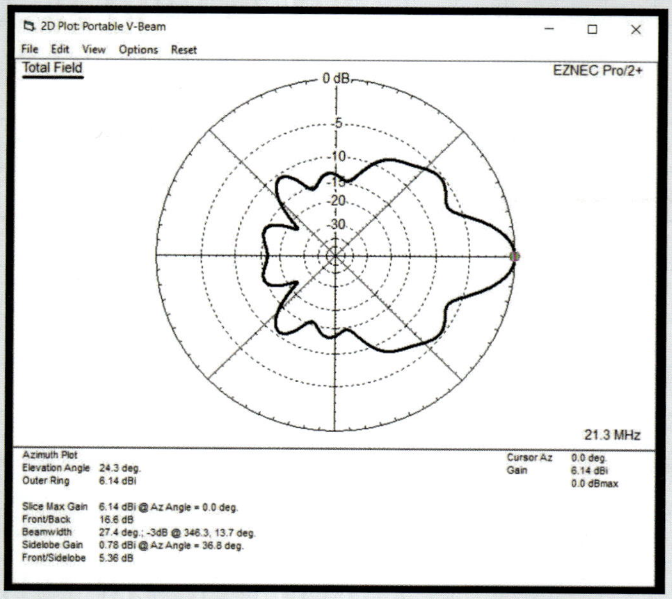
2D Plot: Portable V-Beam
File Edit View Options Reset
Total Field
EZNEC Pro/2+
0 dB
21.3 MHz
Azimuth Plot
Elevation Angle 24.3 deg.
Outer Ring 6.14 dBi
Slice Max Gain 6.14 dBi @ Az Angle = 0.0 deg.
Front/Back 16.6 dB
Beamwidth 27.4 deg.; -3dB @ 346.3, 13.7 deg.
Sidelobe Gain 0.78 dBi @ Az Angle = 36.8 deg.
Front/Sidelobe 5.36 dB
Cursor Az 0.0 deg.
Gain 6.14 dBi
0.0 dBmax

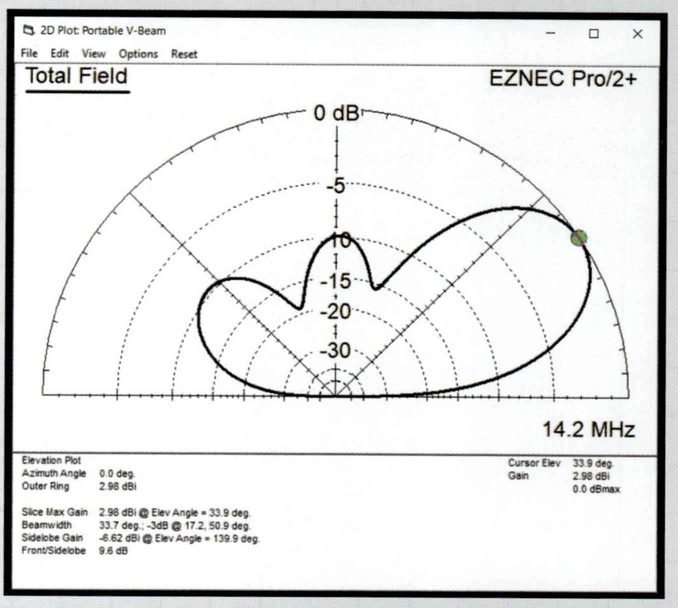
2D Plot: Portable V-Beam
File Edit View Options Reset
Total Field
EZNEC Pro/2+
0 dB
14.2 MHz
Elevation Plot
Azimuth Angle 0.0 deg.
Outer Ring 2.98 dBi
Slice Max Gain 2.98 dBi @ Elev Angle = 33.9 deg.
Beamwidth 33.7 deg.; -3dB @ 17.2, 50.9 deg.
Sidelobe Gain -6.62 dBi @ Elev Angle = 139.9 deg.
Front/Sidelobe 9.6 dB
Cursor Elev 33.9 deg.
Gain 2.98 dBi
0.0 dBmax

I built this antenna using the 12:1 balun and the 600 ohm terminating resistors which I purchased to construct the rhombic and TEFV antennas. Feeding this antenna with 50 ohm coax and terminating the antenna with the resistors gave it incredible directional gain. This configuration also gave me great SWR across all bands, which was a surprise because the research I did on this antenna suggested it required a tuner to achieve good multiband performance. To build it, I attached resistors to PVC pipes and drove the pipes into the ground about 30 meters apart, then hoisted the feed point balun and feed line up on a 25 foot fiberglass mast.

From the mast, each antenna leg, made of #20 AWG wire, ran out to the resistors on the PVC stakes. It wasn't the optimal configuration for this antenna, since the legs sloped considerably, but it did allow me to build it with just one elevated mast and achieve good results. Elevating the legs so that they are completely horizontal would lower the takeoff angle of the transmitted signal and would be better, but even my compromised setup worked great as a DX antenna, especially on 10 meters. It also worked well on 20, 17, 15, and 12 meters.

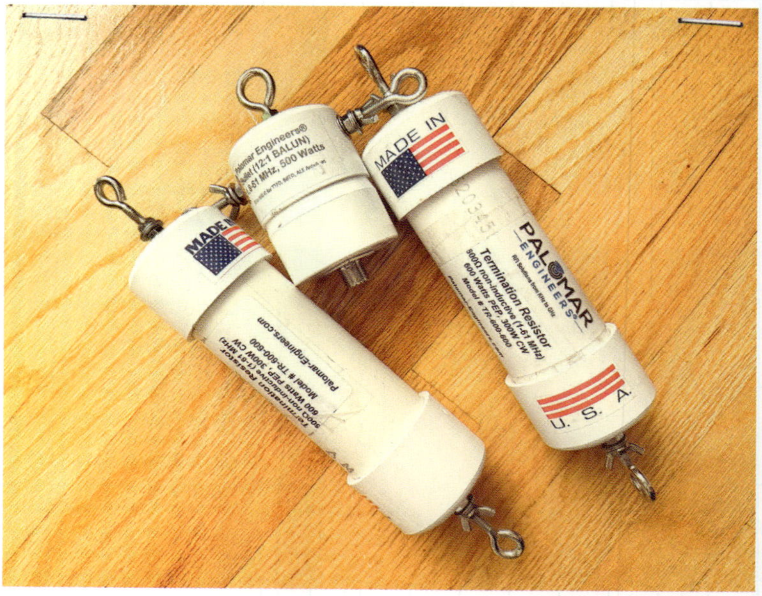

This antenna took some effort to build as a portable setup but was well worth it. If you have the space and the wire to build it, give it a try. Remember, the longer the wire the better, especially if you can run wires around four wavelengths long. Keeping them as high and close to horizontal as possible lowers the takeoff angle, making this a monster for DX.

V-BEAM ANTENNA

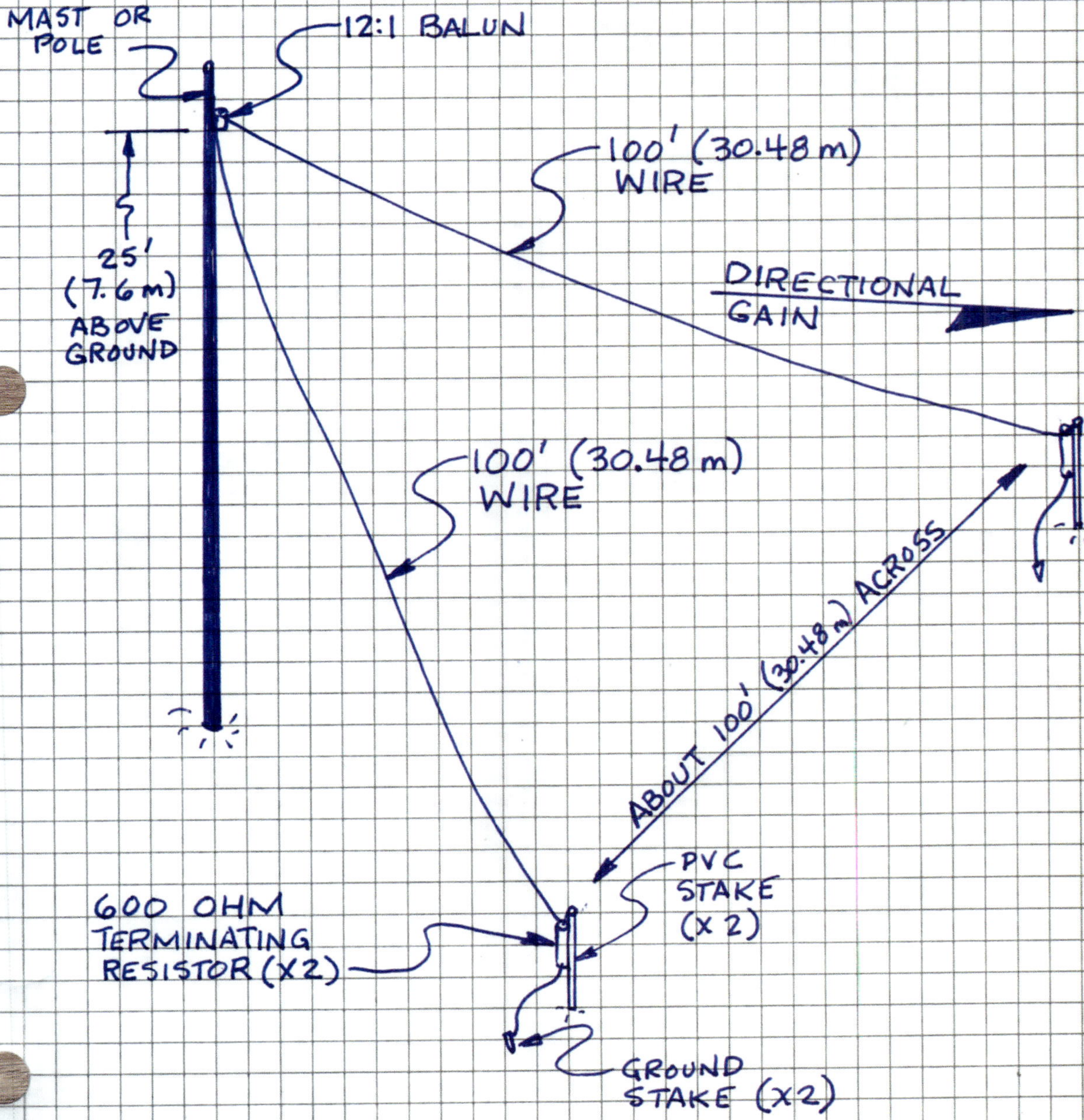

SALTY WALT K4OGO

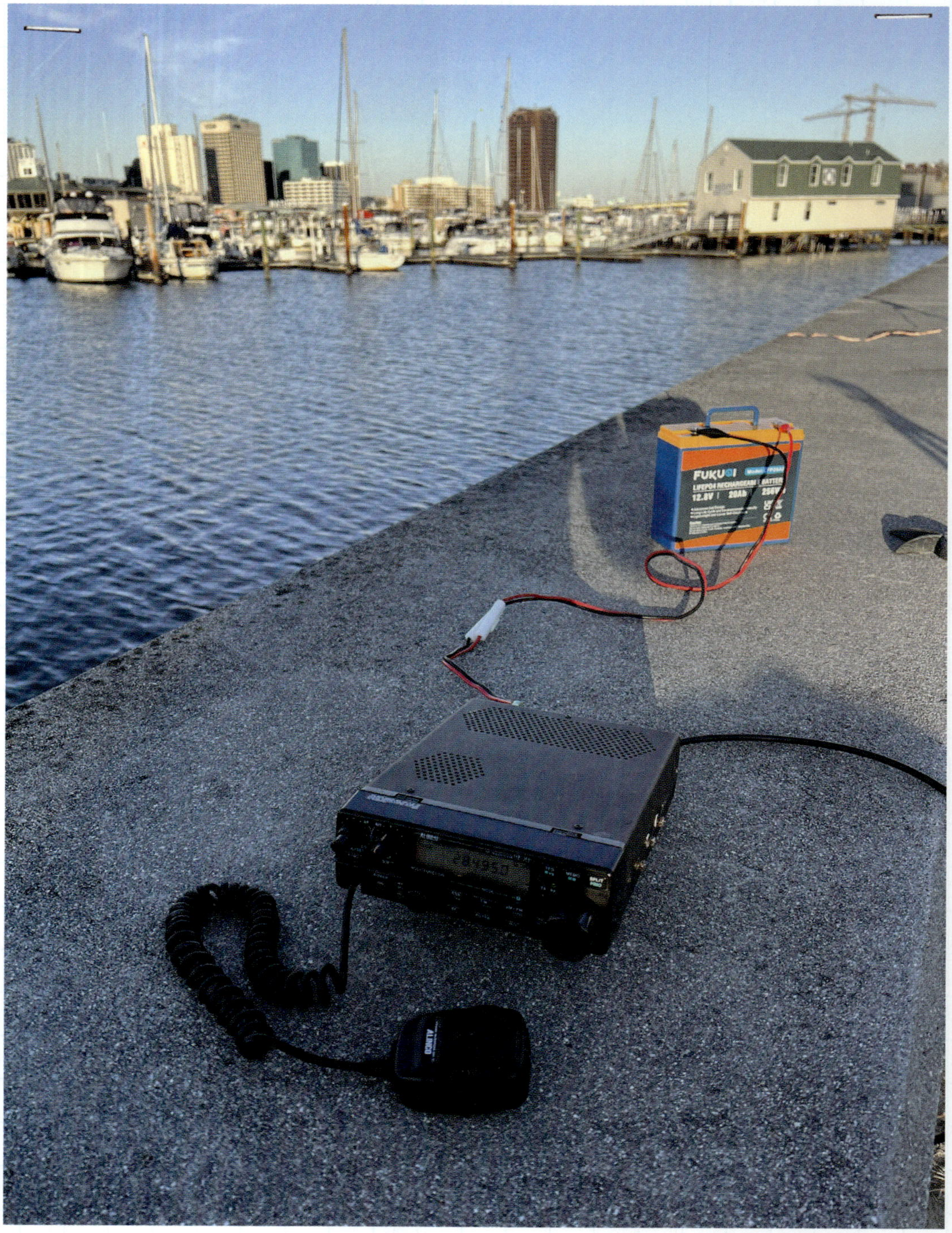

Salty Walt's Portable Rigs

The DX-70

The Alinco **DX-70** is a compact, rugged 100 watt, all-mode transceiver. It's no longer in production, but is available on the used market. It covers the 160 – 10 meter and 6 meter bands, and is about the size of a standard mobile CB radio. This little rig is one of the smallest and lightest options for pumping out 100 watts while out portable. If you can find a good working one on the used market, it's an excellent budget-friendly workhorse.

The Tuned Transmission Line Trap

The tuned transmission line trap (T2LT) antenna is an extremely simple to build, low-cost, portable, vertical dipole antenna built from coaxial cable. Typically, this antenna is used on the 10 meter or 11 meter (CB) band. It requires no ground plane or radials, making it ideal for field use. The heart — and key ingredient — of this antenna is a choke placed at the feed point.

The antenna itself is simply a piece of coax that is a half-wave long with a quarter-wave of the outer shield removed from the top. This antenna has also been called a "flowerpot" antenna because the coax's inner wire rising out of the stripped-clean top end resembles a branch rising out of a flowerpot.

Essentially, this antenna is a vertical dipole and is confined to the higher bands because of the size of the supporting mast required, but it can theoretically be built for any band and erected by sending a line up a tree or some other similarly large support.

A good 1:1 choke at the bottom of the modified piece of coax is critical to keep all the RF energy above it in the form of a dipole.

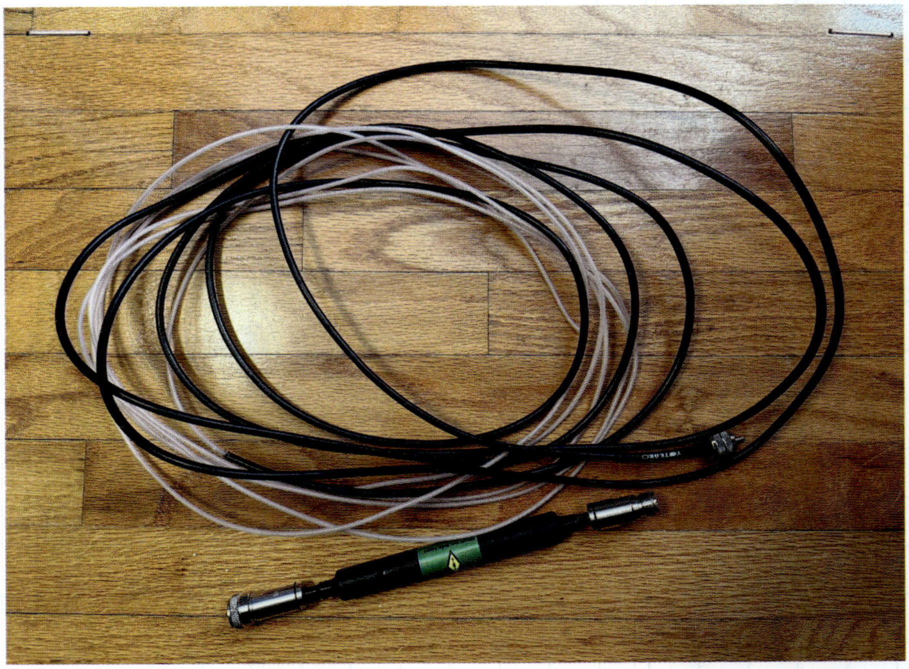

Core Components

Design: It is a sleeve or "flowerpot" antenna where the top has the shield removed, and the bottom uses the outer braid as a radiator.
Performance: Offers a radiation pattern similar to a vertical dipole, providing excellent DX capability.
Choke Alternative: You can also have 5 to 6 turns of coax around a 4 inch (100 mm) PVC pipe placed below the radiator to prevent the feed line from becoming part of the antenna.
Coax: Use 50 ohm coax like RG8X or RG-58.

Typical Specifications

Gain: 4 dBi
Main Lobe Elevation: 8°
Directionality: Omnidirectional in azimuth.

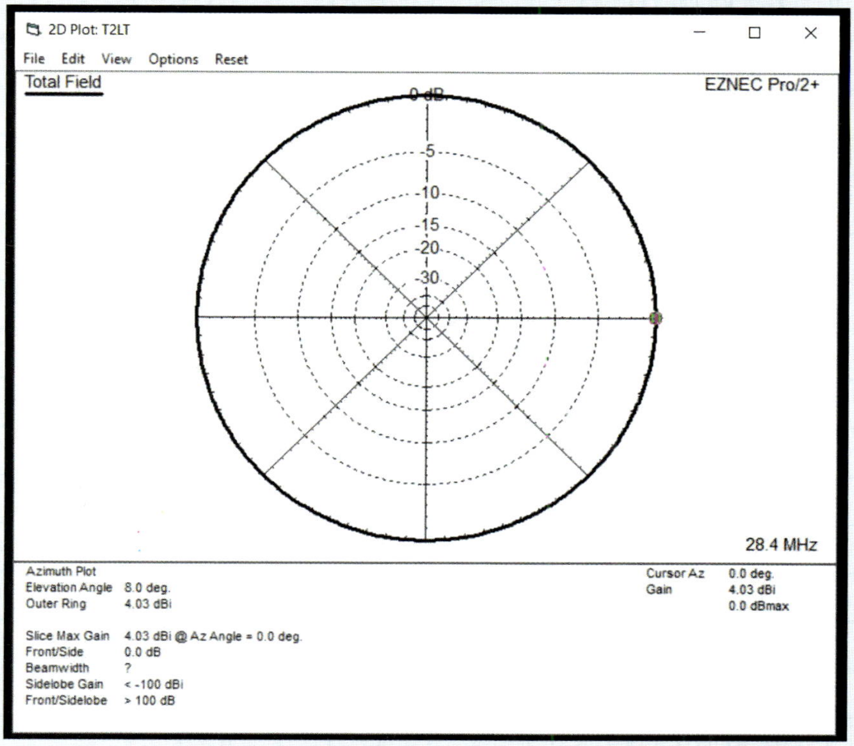

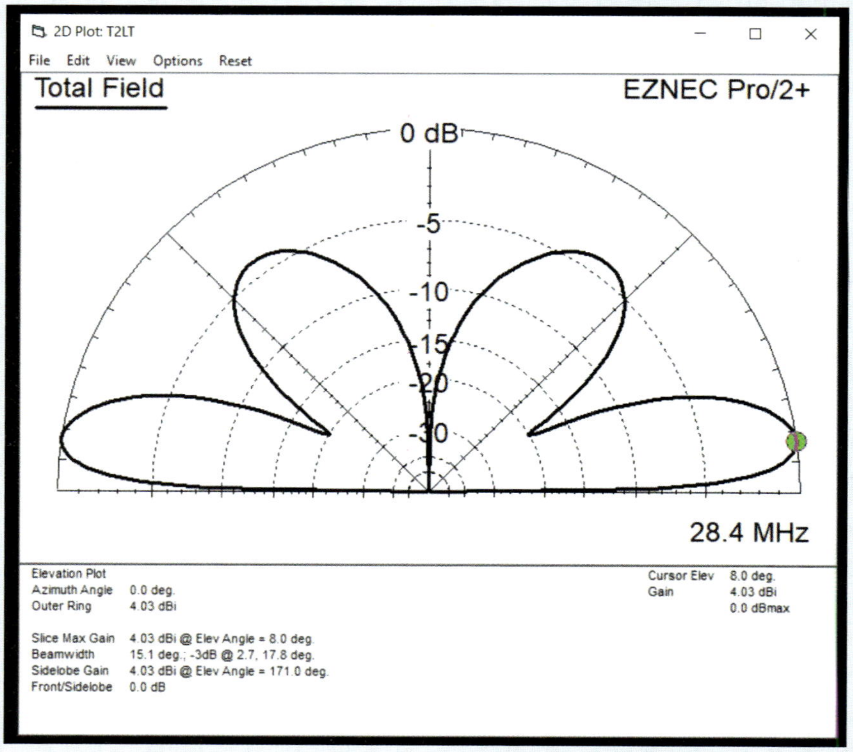

When I built and tested this antenna, I parked my jeep on top of a parking garage at the beach and rolled the front wheel over a flagpole holder. My 10 meter long, fiberglass DX Commander pole fit into the flagpole holder, and I attached a commercially made choke and RG8X coax to the top of the mast. I was completely blown away at the incredible DX I was working with an old piece of coax I had lying around the shack. This quick and easy-to-build antenna was a DX beast! If you have an old piece of coax lying around, don't throw it away! Build this fun and effective antenna.

T2LT ANTENNA

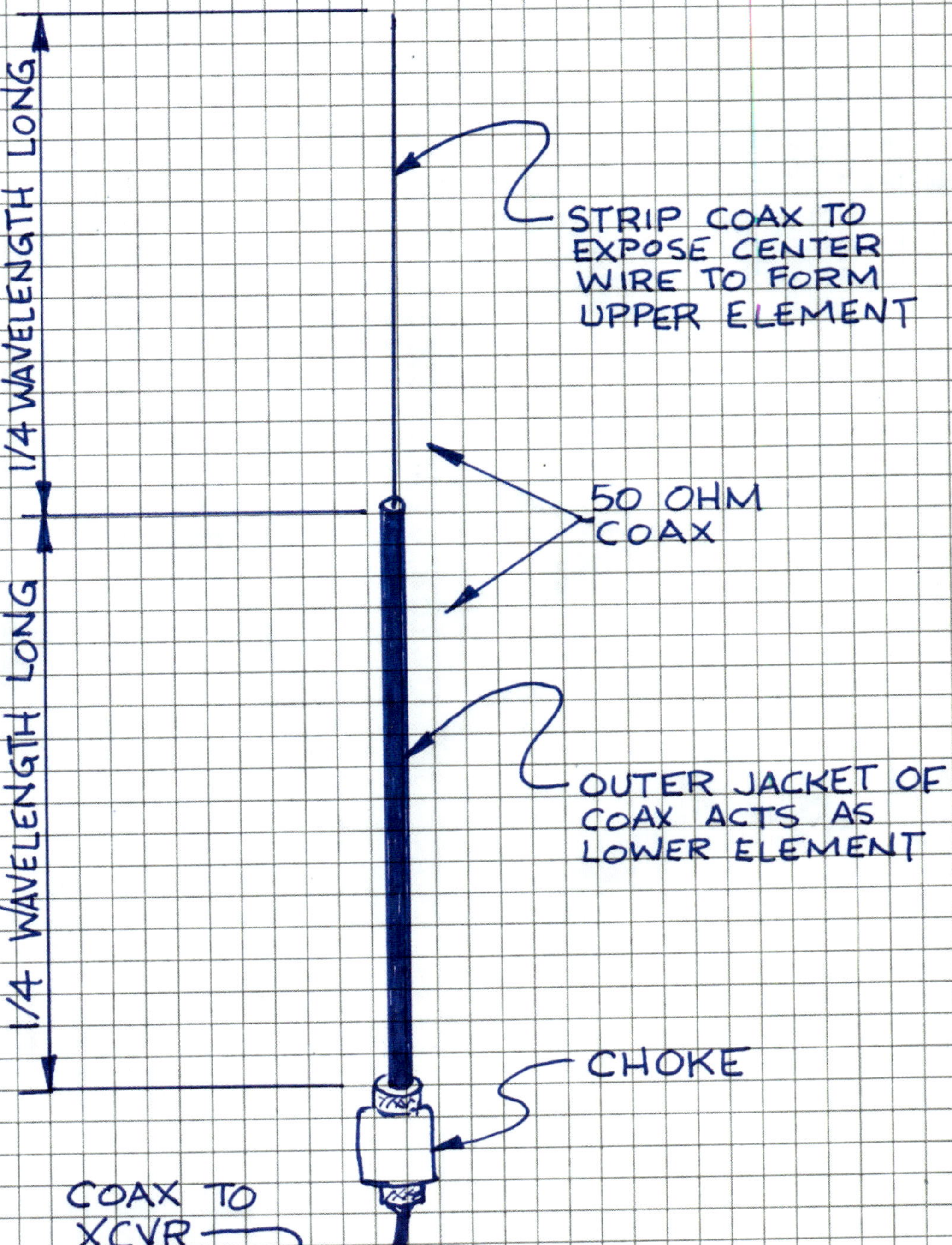

SALTY WALT K4OGO

VOL
RFG
SQL
POWER
VOLUME
100
ATU
X6200
SETTING1
SETTING2
DISPLAY
SYSTEM
MEMORY

Salty Walt's Portable Rigs

The X6200

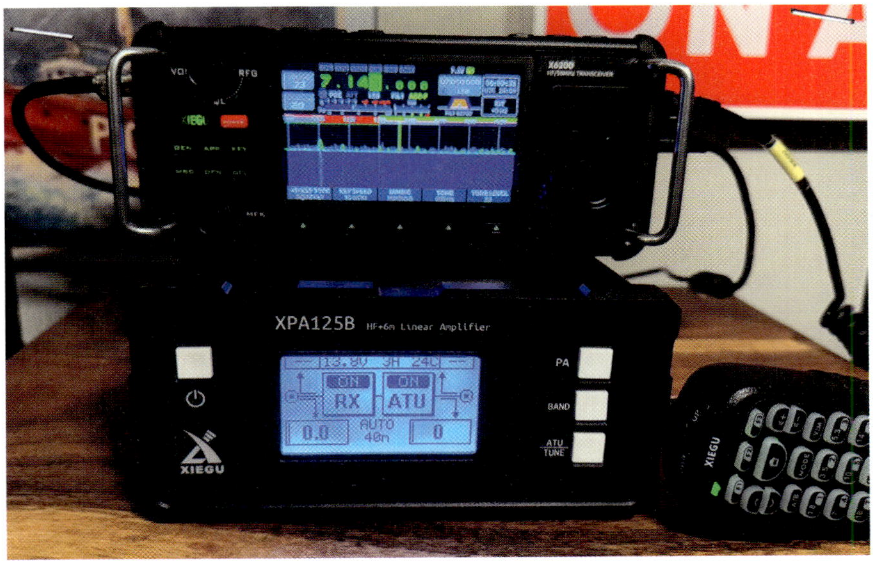

The Xiegu **X6200** is a compact, 8 watt, portable, SDR transceiver (HF through 6 meters) designed for field operation. It features direct sampling, a 4 inch color display, a detachable 3200 mAh battery, a built-in modem for FT8 and digital modes, an automatic antenna tuner, and Bluetooth/WLAN for wireless operation. This is an excellent little QRP rig to use if portable digital modes is your thing. Mine's great for SSB as well. It's a compact rig that feels sturdy and well built.

Next Level "Off the Shelf" Antennas

The ARRL Momobeam

The ARRL Dual-band Momobeam 6/10 Antenna is two antennas in one. It sports a 10 meter Moxon (28 MHz) and 6 meter Yagi (50 MHz) fitted together on a single boom.

Produced exclusively for the ARRL by Momobeam, the antenna is built with aluminum tubes and includes galvanized steel hardware. You can assemble it for portable use with the supplied wing nuts or install it permanently using the additional hardware that is also supplied.

On 10 meters, this design reduces the traditional Yagi by about 30% without requiring traps or other shortening techniques that reduce performance. A Moxon antenna is the genius of Les Moxon, G6XN (SK) and is an extremely effective two-element beam antenna.

On 6 meters, the antenna uses an open-sleeve feed method, coupling the driven element of the 6 meter section by proximity and length to the 10 meter driven element with minimal loss, creating a true two-element Yagi.

The antenna has a power rating of 1500 watts, so it can handle more than any rig you could possibly take out portable could throw at it, but is also a great option for a permanent installation for 10 meters and the 6 meter "magic" band.

This antenna is a great portable option for a day of operating on the higher bands, especially during a contest or on a mid-summer day when 6 meters is open. I attached it to a 5 meter tripod mast that I have without any need for guy wires or securing, although this was on a calm day and I would probably consider securing it a little more on a windy day. The first time I assembled this antenna it required a little time (about an hour), but after that, I disassembled it into three parts that fit in the back of my Jeep for portable outings and quicker deployments. This is a solid antenna and a great way to get out portable with a high-band beam. I made incredible contacts on 10 meters with it and was even able to grab a few contacts on 6 meters when the band was not exactly wide open.

If you want to dominate the portable division of either a 6 or 10 meter contest, this is definitely the antenna to use!

K4OGO
USB
A 14.289.000
B 07.050.000

Salty Walt's Portable Rigs

The Xiegu G106

The Xiegu **G106** is a simple, compact, low-cost, 5W, portable, QRP transceiver with an SDR infrastructure covering the 80 through 10 meter bands. Operating modes are CW, SSB, and AM. It is also a general-coverage receiver that tunes continuously from 550 kHz to 29.7 MHz and also covers the 88 – 108 MHz FM broadcast band. The G106 doesn't include the same options as most other Xiegu transceivers, like an ATU or an SWR analyzer, but the budget price and its solid metal case make it a great, affordable, QRP option. I love mine and enjoy operating on the beach with it. Having the SDR band spectrum on the display of this little radio makes it the best of the budget QRP rigs.

Portable Hexbeams

A Hexbeam antenna is a compact directional antenna popular for HF (6 through 20 meter) DXing. When built for multiple bands, this six-sided antenna has an upside-down umbrella shape. It provides great forward gain and has a high front-to-back ratio, making it ideal for small yards, portable use, or on inexpensive, low masts rather than heavy towers. The antenna simply takes a two-element beam and wraps it around a hexagonal shape. The driven element is made up of two wires that are fed at the center of the hexagon and that extend out to the perimeter. They then run away from each other along the perimeter, forming an M shape, with the top of the M leaving an opening that creates the "front" of the beam. The element ends in a non-conductive spacer on either end, which in turn attaches to the two ends of a reflector wire that forms the back of the hex shape. If you were to turn the antenna on its side, it would look a bit like Pac-Man, if he were hexagonal.

Core Components

Compact Footprint: The elements are folded so it has a much smaller turning radius around 9.5 to 11 feet (about 3 meters) than a standard Yagi or dipole for the same bands.

Low Mounting Requirements: Its light weight and low wind load mean it can often be mounted on inexpensive push-up masts or light towers at heights as low as 20 – 40 feet (6 to 11 meters).

Typical Specifications

Directional Gain: It focuses RF energy in one direction, providing gain and front-to-back performance comparable to a full-sized two-element Yagi.

Multiband Capability: Most off-the-shelf hexbeams cover six bands (20, 17, 15, 12, 10, and 6 meters) using a single coaxial feedline.

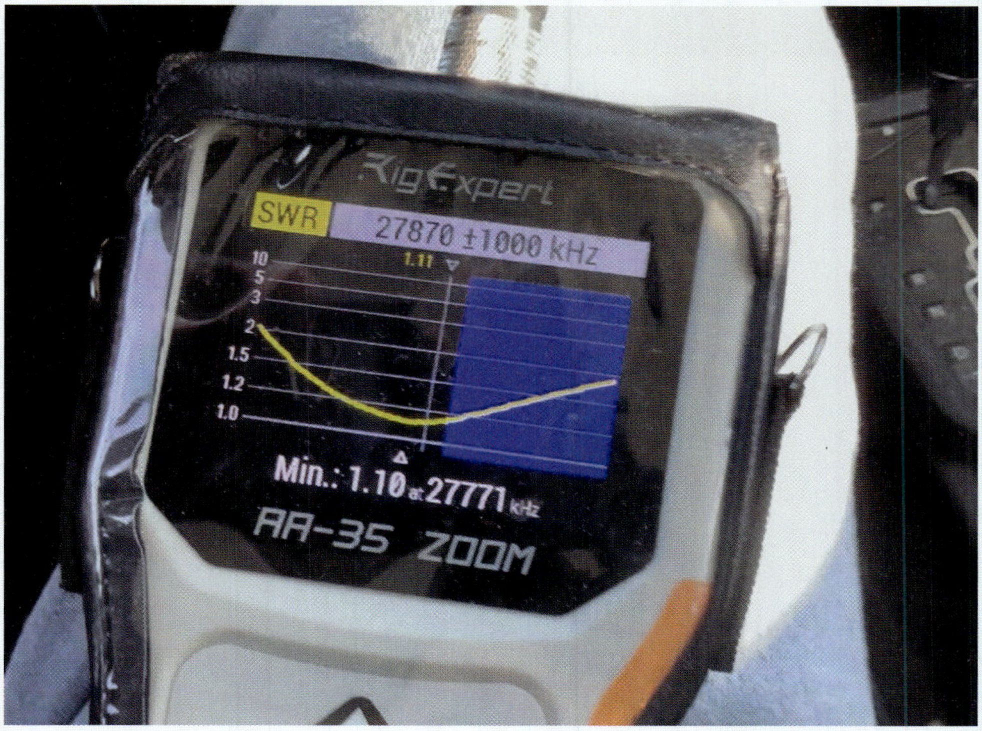

There are many DIY kits and instructions out there for building a hexbeam as well as some great commercial antennas like the BuddiHEX from Buddipole which is a compact portable hexbeam for 20 through 6 meters that comes complete with everything you need to put up a hexbeam in the field. This antenna is wildly popular among Field Day users and portable DXers because of its incredible performance and portability. The antenna is so popular and in high demand that it routinely has a waitlist to acquire.

I found a single band hexbeam built for the 10 meter band that is one of the best portable antennas I've ever used. It's called the ENS POTAbeam, and I discovered it while searching through eBay for antennas. It's much smaller than a typical hexbeam because of its single band design. It's also a fraction of the cost of the multiband beams and much easier to deploy. I was absolutely blown away using this antenna and it quickly turned me into a hexbeam believer.

Single Band Hexbeam Advantages

- RF Gain of up to 10.2 dBi with up to 20 dB front-to-back ratio.
- High gain at low departure angles enhancing long-haul DX.
- Low noise floor due in part to being horizontally polarized; however, directional beam antennas not only enhance forward gain they also nullify unwanted signals and noise from the opposite direction.
- Smaller turning radius (66") compared to traditional beam antenna designs due to the reflected, folded dipole design.

I highly recommend giving a hexbeam a try. Whether you build it yourself or purchase an off-the-shelf version this antenna is a DX Field Day and contest winner.

PA
PTT

Salty Walt's Portable Rigs

The (tr)uSDX

The **(tr)uSDX** is a small 5-band, multimode QRP transceiver that fits in your pocket. It supports CW, LSB, USB, AM, and FM. It covers by default the 80, 60, 40, 30, and 20 meter bands (LO version) but can be purchased prebuilt or in kit form for the 20, 17, 15, 12, and 10 meter bands (Hi version). The **(tr)uSDX** is the successor of the uSDX Project. It's the result of a collaboration between Guido Ten D, PE1NNZ, and Manuel Klaerig, DL2MAN, that created the original uSDX. The design was intended as an easy-to-build (preassembled) kit, but is also available completely built from approved builders. I have the Hi version and worked Australia from Virginia with 5 watts using SSB on 20 meters as my first QSO on this tiny little transceiver. It lacks many bells and whistles and is probably not the rig for a beginner, but for the experienced QRPer, this is a fun, little, inexpensive rig to play with.

The Vertical Moxon

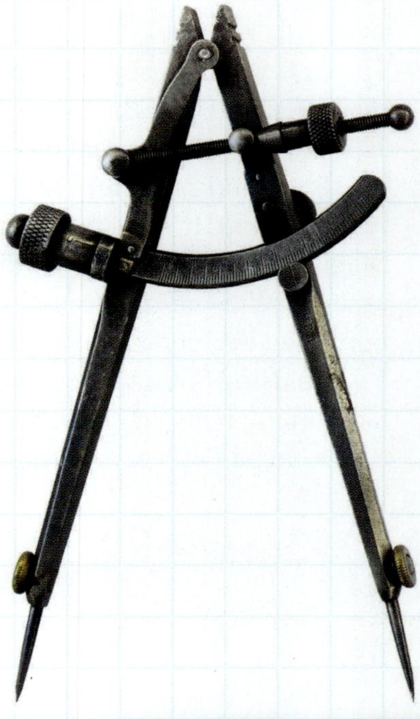

In the previous chapter I spoke about my little 5 watt (tr)uSDX transceiver and how the first contact I made with this tiny radio was an unbelievable SSB QRP conversation with Australia from Virginia. This contact made me more of a believer in the antenna on the other end than the radio I was using or propagation at the time. This was my first of many contacts with my friend Cory, VK4KA, in Brisbane. At the time he was using a Moxon for the 20 meter band on a tower positioned in traditional horizontal fashion. Over time, I traveled to many beaches as well as work ventures to Poland, and he could always hear me no matter where I was at in the world, and I could also hear him blasting in around S9. It gave me "antenna envy," and I wanted a Moxon for 20 meters but just didn't have the space or ability to put one up at my home QTH.

So, what is a Moxon antenna? The Moxon antenna is a rectangular, two-element, directional, parasitic array with elements folded to reduce its size by 30% compared to a Yagi. It is known for high gain, excellent front-to-back ratio, and a 50 ohm feed point requiring no tuner. Typically, it is a DIY antenna for the 10, 6, and 2 meter bands. It is ideal for reducing noise and improving DX performance. It takes its name from the British operator and antenna handbook author Les Moxon, G6XN. Forming a rectangle by taking a two-element beam and folding the ends towards each other, the driven element is slightly shorter than a half-wavelength, and the parasitic reflector behind it is slightly longer than a half-wavelength. The ends of the two are mechanically separated from each other with a spacer, creating a two-element beam bent into a rectangle.

Knowing I can't permanently mount a Moxon antenna at my home QTH, I always dreamed of designing or toying with building a portable one to use on the beach by turning it upright and making it a "vertical Moxon." As this antenna is still rather large to build for the 20 meter band, I just filed it away in my thoughts as a "someday" dream antenna. One day while skipping through YouTube videos I ran across something that excited me greatly. There it was, a commercially built vertical Moxon for 20 meters. Tony (MØTNY) and Dave (MØTPT) from the UK ham radio store Martin Lynch & Sons were out portable with a new antenna, the MyDEL 20 meter vertical Moxon wire antenna. It's a collaboration between ML&S and Pro Antennas in the UK.This antenna was inexpensive, and it was in my cart before the video ended. Never had I wanted an antenna so bad, and to me, this antenna was a dream come true.

Typical Specifications

Lightweight and Portable: quick to set up and pack away.
Simple Installation: Recommended with 2 x 9 meter fiberglass poles.
Power Rating: Handles up to 1000 W PEP
Compact Footprint: Approximately 8 meters vertical height and 3 meters horizontal.
Gain: Around 5 dBi.

This antenna arrived quickly and early one morning I took it out to a local park and assembled it using two DX Commander Expedition poles (10 meters each). I was surprised how easy this was to set up, and then it did not disappoint. This antenna is a DX beast! When you consider its low cost, the time you save in setting up, and how easy it was to purchase, this antenna's a real value. If you're ever out on the Outer Banks of North Carolina or over at Virginia Beach and see a wire rectangle towering above the surf, it will probably be me. I love this antenna.

Salty Walt's Portable Rigs

The RS-918

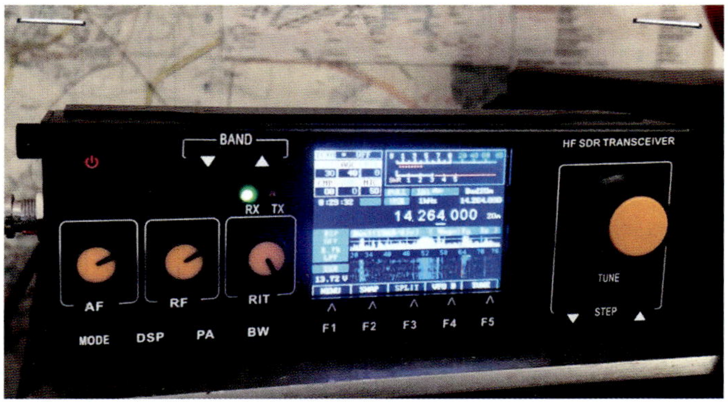

The **RS-918** is a 15 watt QRP transceiver that is a clone of an open source transceiver design, the MØNKA mcHF. This compact transceiver has an internal battery, making it compact and easy to carry. It operates on all bands from 160 to 10 meters. Built-in modes are CW, LSB, USB, AM, FM (narrow and wide), SAM, and FreeDV. If you're looking to get into the digital voice world of ham radio, this transceiver might be what you're looking for. Typically, this radio can be found at a reasonable price, but take caution and shop wisely because this radio is mass produced in the radio clone world. Purchase from a reliable seller online. I had good luck with mine, and it's a joy to use.

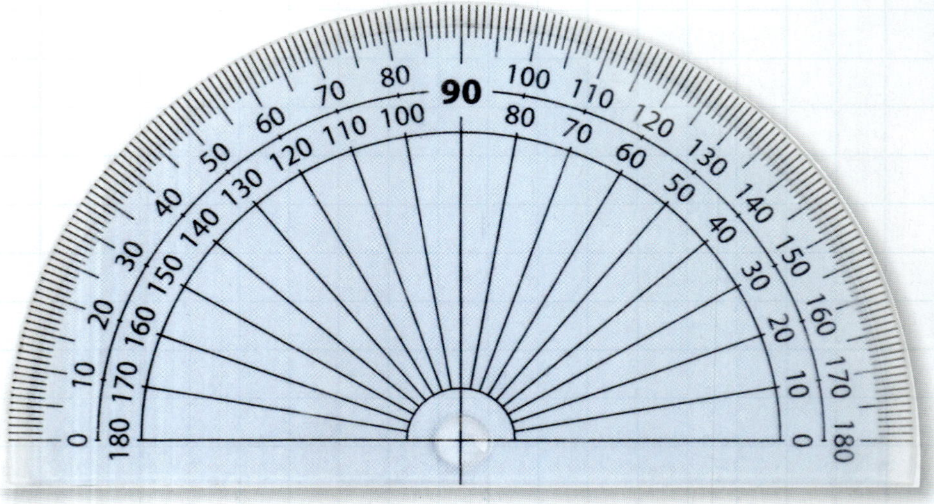

VERTICAL PHASED ARRAYS

In an earlier chapter I showed the co-phasing of two antennas, which is one of the simplest methods for using phasing to form a beam. I did this with two quarter-wave vertical driven elements that, when spaced appropriately, typically a quarter or half wavelength apart and fed in phase with properly calculated cable lengths, formed a beam. Changing the lengths of the feeding cables can take this setup to the next level. For instance, with antennas one half-wavelength apart, equal-length cables cause the antennas to act as one, creating a directional broadside pattern, this is considered "in phase." With antennas moved in to a quarter wavelength, the difference creates a strong bi-directional beam towards the rear and front of the antennas, this is called "90-degrees in phase." If one antenna is fed with a cable half a wavelength longer than the other, shifting the signal nulls and lobes, it creates a figure-8 off the sides. This is called "180-degrees out of phase." Rather than all this cable switching and element moving, the simpler option is to just use an antenna phasing unit.

There are several different antenna phasing systems available for amateur radio, from elaborate systems that rival AM broadcast radio setups to simple and easy-to-build kits. I have a kit-built system from Performance Z Arrays, a manufacturer based in South Carolina. John Gendron, NJ4Z, is the owner and designer of their phasing systems. He uses the Christman method of phasing. Developed by Alan Christman, K3LC, it allows the use of two vertical elements to produce two end-fire patterns and one broadside pattern. What Christman did was compute the point where the voltage magnitude of the feedlines was equal and thus could provide the proper power in the correct timing to produce a 90 degree phase-shift between the elements. With the elements spaced 1/4 wavelength apart, this creates a constructive interference pattern in one direction and a destructive interference patten in the opposite direction, or, in other words, a cardioid pattern in the direction of the constructive interference and a null in the direction of the destructive interference. When compared to a single vertical element, the end-fire pattern will produce up to +3 dB gain with as much as 20 dB in front-to-back ratio in one of two directions. This design allows for the shorting of the delay line, thus feeding both elements in phase and creating a somewhat omnidirectional pattern with a +1 dB to +1.5 dB gain broadside to the array. The array design consists of two equal-length quarter-wave vertical elements and a radial field for each element. The two elements are fed with 84 degree feedlines and a delay line of 71 degrees used to delay signaling to one of the elements.

So, to explain this in simpler terms: When two identical verticals are set up and fed into the phasing unit, you can form and control the beam of your signal by simply turning a knob. For instance, with my Performance Z Array system, I positioned my two identical vertical telescopic whip antennas to face north and south and had them tuned identically on the 20 meter band. With a simple turn of the Direction switch to the left, I had my beam directed north and was getting solid 5-9 reports from the Canadian Maritimes to me in Virginia. Hunting through the bands, I could hear some signals south of me in Florida but could not contact them when I tried. By moving the Direction switch to the far-right side, I was suddenly able to reach those Florida stations, once again with solid 5-9 reports. Twisting the switch to the center position allowed me to talk to my friends in Bermuda, to the east, with ease.

This is a great way to venture into beamforming at a higher level than just manipulating wire. It's also a good way to do it portable or in your backyard without having to put up a big beam on top of a tower. It's definitely worth the effort for working DX.

Salty Walt's Portable Rigs

The Lincoln II+

The President **Lincoln II+** is a 50 watt 10 and 12 meter transceiver and is another example of what operators in the CB world call an "export" radio, because it can be converted for use on 11 meters, and it is widely popular in that market. It has AM, FM, SSB, and CW modes and has a frequency range of 24.890 MHz to 29.700 MHz. Yes, it looks like a CB radio, but it is a great performer at a good price on the 10 and 12 meter bands, especially for portable operations. I have a few President radios, and the **Lincoln II+** is my favorite!

Portable Magnetic Loops

I have a love/hate relationship with magnetic loop antennas or "mag loops" as they're commonly called. I've found that they work but take a bit of learning and skillset development to use. I've also found that not all magnetic loops are equal: finding and using a good one is key.

So, just what is a mag loop? Magnetic loop antennas are compact antennas that use a circular conductor and a tuning capacitor to create a resonant circuit. Unlike traditional wire antennas that respond to electric fields, these respond primarily to the magnetic component of radio waves. This makes them much quieter in noisy environments.

Core Components

Compact Size: Usually around 3 feet (1 meter) in diameter. Some claim they are great for apartments, balconies, and HOAs.
Low Mounting Height: Unlike other antennas that need to be high off the ground, a magnetic loop can perform well just above the ground.

Typical Specifications

High Selectivity (High Q): They act as a sharp filter, only receiving and transmitting in a very narrow bandwidth which helps eliminate nearby noise and channel interference.
Directional Nulls: They feature a "figure-eight" radiation pattern with deep nulls. By rotating the loop, you can physically "null out" local noise and can beam your signal somewhat.

OK, I know this all sounds great, but magnetic loops also have some limitations and drawbacks. One limitation, for example, is narrow bandwidth. Because mag loops are so selective, you must retune the antenna every time you move the VFO even a few kilohertz on the radio. This can be extremely frustrating if you are portable and are hunting signals. Another limitation, or really a major drawback, is the existence of a voltage hazard. During transmission the tuning capacitor can develop high voltage, even at low power, making them dangerous to touch while in use. Another drawback is the power limits mag loops impose. Standard portable loops are often limited to 10 to 20 watts PEP, although there are a few 100 watt options out there.

As a newer ham, I had the romantic idea that a mag loop would be my hotel balcony travel antenna end-all and be-all. That idea died quickly. Using a mag loop takes some practice and patience. I've found that the best way to tune a mag loop is to find the spot where you are hunting or want to call CQ and then turn your radio's RF gain all the way up. Turn the magnetic loop's variable capacitor until you hear your radio's noise level reach its peak. This is typically the tuned spot, although you still may be off a bit, and as soon as you remove your hand from the antenna's tuning capacitor you might hear the noise level drop. This is because your body is also coupling to the antenna. So, you can see, learning to tune to the exact spot may be frustrating and take a bit of practice. I've also just found what I thought is a good tuning and then hit the radio's internal tuner but only if I'm calling CQ. This isn't a practice to follow when hunting signals.

As I stated, finding a good magnetic loop antenna is important. I have purchased and played around with several and have whiled away many frustrated hours. One mag loop I do have and enjoy is the AlexLoop HamPack Portable Antenna. It worked great for me sitting on the beach on all bands from 40 to 10 meters and can handle 25 watts. It folds up nicely and stores away in a small bag. I really like this antenna, but once again my advantage was sitting on a beach in an ideal spot to operate. I could have operated with any antenna and gotten good results.

There are some good options for magnetic loop antennas out there, and I would highly encourage anyone to try one because they can be enjoyable, and figuring out how to make one work is a great learning experience.

Salty Walt's Portable Rigs

The SR-955HPC+

Another "export" style radio from the CB world that I own and love is the Stryker **SR-955HPC+.** This is a high-performance 10 meter transceiver that boasts over 100 watts PEP output in AM, FM, and SSB modes. It features advanced digital noise reduction. On the bench, mine tested out at 110 watts on 10 meters and is a great small option to get that type of power when operating mobile or portable. This is a great radio to permanently install as a mobile or to just take out potable. It's got punch and sounds great!

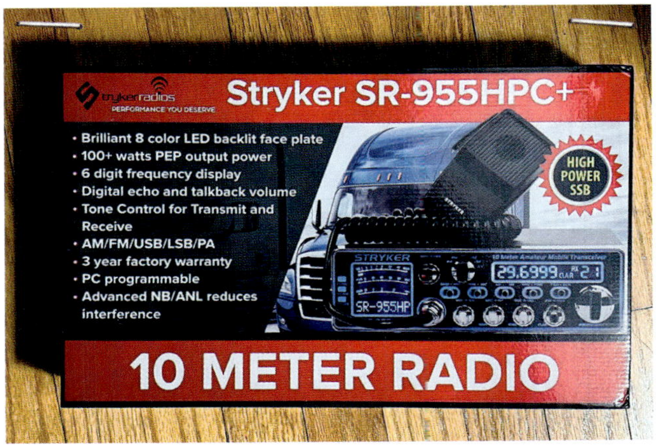

Variations on the Telescopic Whip

Telescopic whip antennas provide an easy way to experiment with and build a variety of antenna setups. They come in several sizes, but the most common is the 17 foot or 5.6 meter long antenna, which is easy to find with either M10 or Imperial (USA) 3/8"-24 threads. You tune these antennas for the higher bands (20 through 6 meters) by extending or shortening the telescoping sections. But these antennas aren't limited to the higher bands, they're also a popular option for the lower bands, when the antenna is electrically lengthened by inserting a coil. Examples of great coil-fed, telescopic whip antennas include the Wolf River Coil or Slidewinder DX.

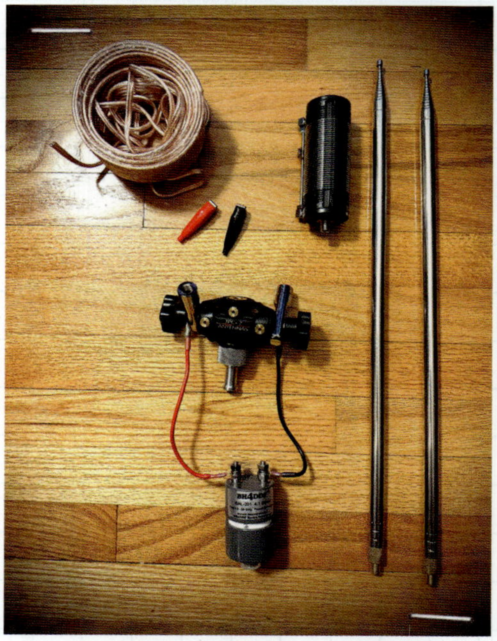

There are several antenna packages you can purchase that come complete with coils, mounting spikes, and other accessories. These packages are convenient, easy to carry, perform well, and a bunch of different brands and manufacturers sell them. Some common ones are the Buddipole "Buddistick," the PAC-12, JPC-12, Radioddity HF-009 and HF-010, the MC-750, JPC-350, and on and on. There are so many different types and packages that trying to nail down one perfect setup is impossible.

These telescopic-type antenna systems also come in variations that allow you to set them up as portable dipoles or positive V dipoles for all the HF bands using either a mount and tripod for the higher HF bands or coils on the lower HF bands. One system in particular, the JPC-7 antenna, adds a dipole center feed point, a coil, and a telescopic whip to the JPC-12.

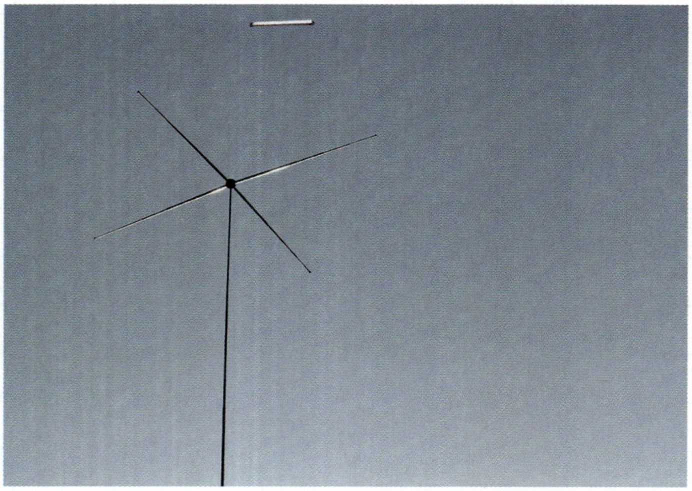

These types of antennas also have several neat add-on accessories like radial plates for additional ground radials or capacitance hat kits. It was with the "cap hat" add-on that I made my first portable contact in the greyline from North Carolina to Australia on 40 meters. That's something that I regularly do on 20 meters but not 40. This was a thrill, and I'm convinced it was the cap hat over the JPC-12 with a coil that made it happen.

I've gathered a collection of telescopic whip parts, and I think this is a great way to build many different antennas. I even used a couple telescopic whips to build my co-phased setups as well as my phased array vertical setup. Going even farther, I took the JPC-7 dipole system and used just one coil and positioned the dipole in a V. Then I clipped a line of #16 AWG wire across from the tips of the top of the whips forming a Delta Loop that was tunable from 20 to 10 meters. There are so many more out-of-the-box type things you can do with this type of antenna.

If cutting wires and erecting a tall mast are not your thing, playing with telescopic whips just might be the answer to getting out and experimenting with different antenna configurations portable. Get out and give it a try!

Salty Walt's Portable Rigs

The Scout 555

One of the last great American manufacturers of amateur radio equipment was Ten-Tec. Founded in 1968 and based in Tennessee, during their heyday this company made some great radios. One compact rig they produced from 1993 through the early 2000s was the Ten-Tec **Scout 555**. It remains a popular portable rig for field operations. It uses a unique plug-in module system and each band (80, 40, 30, 20, 17, 15, 12, and 10 meters) requires a separate sliding module. It is a quiet and clean receiver and has a built-in Iambic Keyer for CW that is adjustable from 5 to 50 WPM. Its modes are SSB and CW and its power is adjustable from 5 to 50 watts (PEP). It's an oldie but goodie and can still be found in great working condition on the used market. I absolutely love my Scout!

Conclusion

Get Out There!

If this book accomplishes one thing, I hope it encourages you to go outside, put up an antenna, and have some fun. Having fun is the goal of all my ham radio adventures. But beyond that, I hope this book also helps you hone your antenna assembly skills and gives you just enough theory to whet your appetite for more. My first book, *Salty Walt's Portable Antenna Sketchbook,* focused on setting up simple vertical and horizontal wire antennas. This book picks up where that one left off. The antennas in it are more complicated—but still achievable!—and as you build them, you'll learn to form beams using multiple wires and more complicated configurations. You'll experiment with signal direction and intensity. Those skills are the "Next Level" that reading about and assembling these projects should help you achieve.

I've had fun building these antennas, and I even learned a few things along the way. I hope you do too, but most of all, I hope this book inspires you to grab your radio, get out there, and start building antennas. I look forward to catching you on the air soon.

73, and "stay salty" my friends!

"Salty" Walt Hudson, K4OGO